University of Toronto Studies
Biological Series, No. 43

FLUCTUATIONS IN THE NUMBERS OF THE VARYING HARE
(LEPUS AMERICANUS)

By
D. A. MacLULICH
Department of Biology, University of Toronto

VELUT ARBOR ÆVO

TORONTO
THE UNIVERSITY OF TORONTO PRESS
1937

SCHOLARLY REPRINT SERIES
ISBN 0-8020-7039-6
ISBN 978-1-4875-8178-7 (paper)
LC 38-22317

PRINTED IN CANADA
Reprinted in 2018

CONTENTS

FLUCTUATIONS IN THE NUMBERS OF THE VARYING HARE (*LEPUS AMERICANUS*)

INTRODUCTION

The varying hare, *Lepus americanus* Erxleben (1777), has become an almost classic example of an animal that undergoes periodic fluctuations in numbers. This phenomenon has in recent years attracted a great deal of attention, and appears to be of considerable biological importance. Clark (1936) has recently given a summary of the growth of our knowledge of this subject. Since the case of the hare is such common knowledge, and since hares comprise one of the very large and important populations in the economy of nature, it was considered important to make an investigation of it.

PREVIOUS WORK

It has long been a matter of common knowledge that the population of hares varies widely from time to time; Richardson (1829) noted that hares were sometimes abundant, sometimes scarce, and that they suffered from epidemics. Many explorers and naturalists have reported particular times of great abundance without recognizing the periodic nature of the fluctuations (Hind, 1860; Macoun, 1882; MacFarlane, 1890; and others). Poland (1892) also failed to note the periodicity, although he published fur returns that showed it; he simply mentioned the fact that hares were abundant in some years and not in others. The first real addition to our ideas since Richardson's day was made by MacFarlane (1905), who recognized the ten year periodicity of the fluctuations. The contributions of Seton (1909, 1928) and Hewitt (1921) will be mentioned in discussing the use of fur returns as records of animal abundance.

More recently the problem was discussed at the Matamek Conference on Biological Cycles (Huntington, 1932; Elton, 1933) which was called together by Mr. Copley Amory. Elton (1924, 1933b, 1933, 1934, 1935) and the Bureau of Animal Populations, of which he is Director, have been studying the

5

Canadian hare cycle from the records of the Hudson's Bay Company and by a questionnaire maintained by the National Parks Branch of Canada. Leopold (1931, 1933) has recorded the history of the hare cycle in Wisconsin, and discussed its bearing on game management. Green and Shillinger (1931, 1932, 1934, 1935) have been conducting an annual census on an area in Minnesota by trapping hares alive in large numbers and tagging them, and studying their diseases.

In the present investigation two preliminary announcements (MacLulich 1935, 1936a) and a paper on the relation of sunspots to animal abundance have appeared (1936c).

Record of Fluctuations in Abundance

Four sources of information on the abundance of hares were available, namely:

    (1) Records of furs taken by trappers; the fur returns of the Hudson's Bay Company were used and some figures from the Dominion Bureau of Statistics.

    (2) Statements in the literature.

    (3) Questionnaires.

    (4) Field work.

The methods of handling the material of the first three of the above sources will be described first, followed by a statement of the history of the cycle as disclosed by that material. The field work will be dealt with last.

Material and Method of Compiling History of Fluctuations

Fur returns

The hides of the varying hare are of commercial value and the Hudson's Bay Company has shipped large numbers from Canada to their London market. The records of the numbers of hare pelts handled by the Company have been used as an index of the abundance of hares in Canada by several writers mentioned below and again in the present study.

In connection with this use of the fur returns it is necessary to review the literature critically and discuss some points about the method of using them. This review is included in the writer's paper on sunspots and the abundance of animals (1936c). The only published tables of figures of fur returns of the Company were obtained by Poland (1892), and include from the year 1752 to 1890. His figures seem reliable as he was careful to state what they represented (quoted below) and in footnotes mentioned when any shipment had been delayed or lost.

MacFarlane (1905, 1908) gives Hudson's Bay Company figures for a small number of scattered years to indicate the fluctuations in numbers of many fur-bearing animals. There must be some unexplained features about some of these figures, as certain of them do not agree at all with those obtained from the Company by Poland, Seton, and Hewitt.

Seton (1911) wrote: "Through the courtesy of its officials I have secured the Company's returns for the 85 years—1821-1905 inclusive"; and he published graphs of the numbers of pelts with considerable discussion of the fluctuations. The graphs, except in the case of the rabbit, show points for the years 1906 to 1908 in addition to those mentioned above. The numbers of pelts agree with those of Poland's book as closely as may be read off the graphs. These plates were copied in his "Lives of Game Animals" (1925-8).

The fourth independent publication of the fur catch was by Hewitt (1921), who said: "Through the kindness of Mr. W. H. Bacon, late fur commissioner of the Hudson's Bay Company, I have been able to obtain the fur returns of that Company covering a long period of years, from 1821 to 1914." These figures were presented in the form of graphs and the cycles analytically discussed in much detail. Several corrections are necessary but unfortunately not mentioned in Hewitt's book. On the rabbit graph the vertical scale of numbers is only one-tenth of the values given by Poland (1892) and Seton (1911); therefore it is thought that the fur figure 15,000, for instance, should be read as 150,000. In the graph of lynx numbers the values have all been plotted over the year *after* that for which they were quoted by Poland and Seton. On this curve the figure for the year 1868 is higher than that of Poland and Seton for 1867 by 10,000, causing a one year difference in the year of peak abundance. The lynx figures are mentioned because they are used later in this study. Other writers have copied their fur figures from Seton or Hewitt.

In recent years "... The Hudson's Bay Company has made available for study by Mr. Charles Elton its records covering fur catches in Canada for many years past and has provided facilities in the Company's organization for the collection of statistical material bearing on cyclical fluctuations in animal life" (Huntington, 1932; Elton, 1933a).

The fur returns, as published in graph form by Seton (1911, 1928), Hewitt (1921), and all who have used their figures to portray the changes in abundance of animals, are plotted over the year of sale and office return.   This is proven by the fact that the figures in Poland's book (1892) agree with the values of Seton (1911) for the same years.   Poland stated that the Hudson's Bay Company quantities are "those that are imported towards the end of the previous year, excepting those shipments which are delayed by the ice to the north of Hudson's Bay; these do not arrive till the year after".   There is no ambiguity about this statement, which means that fur reported for, say 1880, was sold in the winter and spring of that year in the London fur sales.   It had been sorted and graded, shipped to England by boat out of Hudson bay in the late summer and autumn of the previous year, and had been accepted in trade from the Indians, Eskimoes, and white trappers during the spring and summer of that year and trapped in the winter of 1878-9.   That is, fur reported for a particular year represents the abundance of the animal at the end of the summer season of biological production of *two years* previous.   Elton (1924) said: " ... the Hudson's Bay Company fur returns for any year include the catch of the winter before; in fact that is the main item, *i.e.* a maximum of skins in 1907 means many foxes caught in the winter 1906-7, and in the spring.   Such abundance will, of course, be the result of the year 1906."   In the discussion above it has been shown that the 1907 figure referred to the abundance of animals as "a result of the year" 1905 rather than 1906.   In regard to a graph of the lynx returns for the Mackenzie river district alone, Elton's statement (1933b), "Dates refer to old 'outfit' years, *i.e.*, the years of biological production and not to the year of office return or sale", is unquestionably correct, as the figures were compiled largely from trading post account books.

As Poland's book (1892) is not generally available, it is worth quoting some of the information contained in it. "These vessels make one voyage a year, either from Hudson's Bay or the North-West coast of America. The York Fort ship arrives in London in September; the Moose River ship at the end of July or beginning of August; and the Vancouver Island at the end of November or December. Many shipments are, however, now made by the great steamship lines from Montreal, the transit by steam being more expeditious. Some goods from the North-West district are sent through the West Indies, the passage by Cape Horn being only used for the bulkier and less valuable goods." The record of shipments lost or delayed was given as follows in footnotes. (Abbreviations: M.R. means the shipment from Moose river, E.M. the shipment from East Main river, and Y.F. that from York Fort or Factory.) "1833, Y.F., M.R. not arrived this year. 1834 including Y.F., M.R. of 1833. 1836, Y.F. not arrived this year. 1837, including Y.F. of 1836. 1864, M.R., E.M. not arrived this year. 1865, including M.R., E.M. of 1864. 1873, M.R., E.M. not arrived this year. 1874, including M.R., E.M. of 1873. 1884, M.R., E.M. not arrived this year. 1885, including M.R., E.M. of 1884, no M.R., E.M. of 1885, lost at sea. 1886, part of Y.F. not imported, Cam Owens wrecked. 1887, with part of Y.F. 1886. 1890, part of Y.F. not imported."

Still quoting Poland: "The fur sales of this Company take place at the commencement of the year; the Beaver, Musquash, and American Rabbit in January; and all other furs in March. Up to 1878 the North-West goods were sold in September; in 1880-82 they were sold in July; since that date they have been included in the March sales."

In recent years fur has been taken out quickly by the railway and steamships, although one shipload is still sent each year from Moose Factory through Hudson strait (Dominion Bureau of Statistics, 1927). The writer does not know whether the practice of selling Canadian fur at autumn fur sales has been revived, but such recent developments would not affect the fur returns used in this paper. A number of the Company posts have been abandoned as they ceased to pay their expenses.   Free traders are taking an increasing part of the fur catch and Canadian furs are being sold extensively in St. Louis, Winnipeg, Edmonton, and Montreal (Dominion Bureau of Statistics).   Starting with 1920-1 the Dominion Bureau of Statistics at Ottawa has been publishing the numbers of pelts of various fur-bearing animals "taken in Canada, years ended June 30", in the "Canada Year Book", compiled from reports required of fur traders.

The part of the history of the abundance from 1788 to 1820 (see fig. 3) has not previously been published in graph form, and, although suggestive, it is too irregular to demonstrate much about the abundance of rabbits.   The gap from 1825 to 1843 occurs in all of the source books, and the extremely high total of 285,607 skins for 1843 probably includes an accumulation from at least a part of the above gap.   Poland (1892) wrote: "Quantities are nevertheless subject to a certain degree to the demand. . . . The Indians, on the other hand, trap all sorts of fur-bearing animals, and refuse to do business with a collector if he will not buy all the kinds." Many rabbit skins were utilized by the Indians and at the posts for blankets and clothing and never reached the fur market.   MacFarlane (1905, 1908) said: "The Hudson's Bay Company does not trade rabbit skins in the interior, but from the posts situated on the shores of Hudson Bay they annually export to England many thousands", *i.e.*, the area for the Hudson's Bay Company rabbit fur returns is only the part of northern Canada draining into Hudson bay.   For the purposes of this paper, a peak of abundance carries the idea that it marks the end of the period of abundance.   For calculations based on the hare graph, the year 1864 was therefore used as a peak year, although the figure for the previous year was slightly higher.

Literature

Records of abundance of hares occurring in the literature have been incorporated into the compilation of the questionnaire replies, to be described below.   The references from which information has been abstracted are included in the list of literature cited, and marked with an asterisk if that was the only use of the reference.

Questionnaires

On the questionnaires were provided spaces for recording whether hares had increased or decreased during the previous year and whether they were abundant or scarce, if either. The questionnaires were sent out in January and referred to the year from the previous spring to include that winter, thus corresponding to the Hudson's Bay Company "outfits" and Elton's questionnaires. The data were transferred to filing cards for convenient reference and preservation.

For Ontario the following method has been used. On a map of Ontario a grid of squares fifteen miles to a side was marked so that one set of alternate lines, *i.e.*, thirty mile squares, coincided approximately with the grid used by Elton. For a close study of conditions in Ontario the thirty mile squares were too coarse, and the fifteen mile units were substituted. It would have been desirable to follow the method already described by Elton (1933, 1934, 1935), but certain changes seemed to be valuable enough to warrant modifying his method. The basic data are given in figures 1 and 2, in such form that it may be worked over in the exact way Elton handled his material by anyone interested. The squares have been labelled by numbers across the columns of the map grid (see figure 1) from west to east, and by letters from south to north.

The conventions used by Elton in regard to the areas covered by observers have been modified somewhat, chiefly as a result of the writer's knowledge of topographic and social conditions in parts of Ontario and personal acquaintance with many of the observers. Elton's conventions will be stated, each followed by any changes that have been found expedient.

(1) If the observer stated the area he referred to, that was accepted. "Anything more than a hundred miles across is considered as doubtful unless there is good reason to suppose that the observations are reliable." In many cases one or two of the fifteen mile squares completely covered the area.

(2) "Where the observer gives no details but simply mentions the name of a place, a convention is adopted of drawing a circle of 10 miles radius (*i.e.*, 20 miles across) round the place. In most instances this twenty mile

circle is found to be smaller than the average areas of other observers, and therefore an underestimate on the safe side." (Elton 1933b). The twenty mile circle has

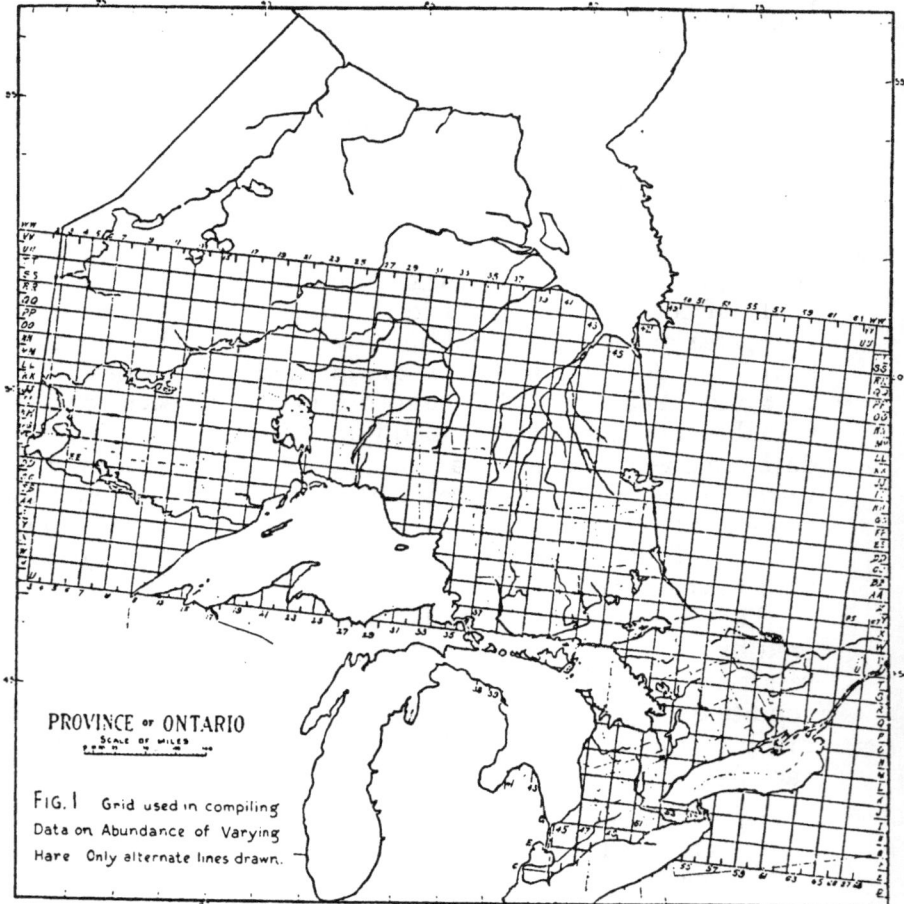

PROVINCE of ONTARIO
SCALE OF MILES

FIG. I Grid used in compiling Data on Abundance of Varying Hare Only alternate lines drawn.

FIGURE 1

been replaced by allotting a total of four of the fifteen mile squares to such an observer, subject to the new rule No. (5).

(3) "Areas . . . are . . . counted as covering a square even if they just overlap a part of it." This has been re-

placed by the modification of No. (2) above, and by No. (5) below.

(4) "Each Hudson's Bay Company post is given a circle with a fifty mile radius (100 miles diameter)." Few of their reports have been available to this study.

(5) Where the writer knew the observer and/or the local situation so as to be able to make a more accurate statement of the area covered than would be given by the above conventions, that knowledge was applied.

Instead of plotting each reported area on a map for the year, they were each looked up on the provincial map with the grid of fifteen mile squares superimposed and recorded in red ink on the filing card. The abundance reports were then tallied in a *tabulation*, for each square of the grid they overlapped. Instead of preserving only the statement, "increasing" or "decreasing", the following categories were used: Increased, Abundant (if stated), Decreased, Scarce (if stated). This basic information is given in figure 2. The average consensus of opinion, based on the majority of the reports for a square, was written in a summary column for each year in the above table. In cases where there was an even division of opinion, no summary was inserted and the square was left blank on the map. The summaries were recorded by the same symbols as were used on the map mentioned below. The allotment of areas in terms of the grid and the handling of any unusual remarks on the questionnaires are the only parts of the whole process that would have to be done by a research worker; the rest could all be done by a careful office helper, if one were available. This process of arriving at one average statement for each square was the major deviation from Elton's method. Elton's method may be criticized for not digesting the material. If there were ten reports of increase and only one of decrease for a square, his method would show the square marked for increasing on one map and for decreasing on the other; the single report would receive as much emphasis and credence as the ten. The ten would probably be correct for the square or for the greater part of its area at any rate, but that would be ignored and hidden by his method.

Only one map, instead of two, was made for each year, and

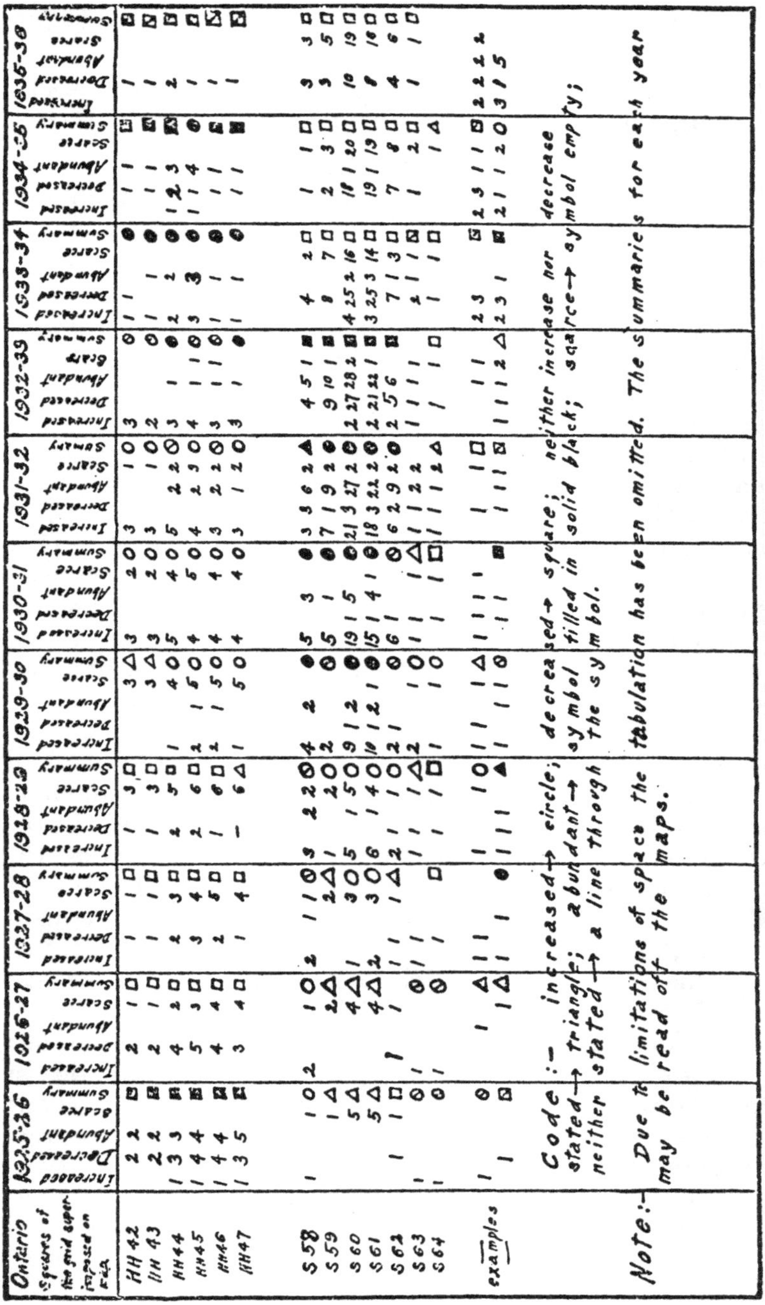

FIGURE 2.—Example of method of tabulation and summarizing.

that showed the average concensus of opinion for each square of the grid, by a single symbol in the centre of the area of the square (see legend on map). The grid lines were not marked, as they would unnecessarily fill up the map.

The graphs used are based on the numbers of squares on which hares were abundant, expressed as a percentage of the total numbers of squares for which reports were available in the province or zone. They show the percentage of the area over which hares were at or near their peak of abundance. A fainter line shows the percentage of the area on which hares increased and another line the area on which they decreased. The greater part of the increase or decrease takes place within two or three years, so that during about half of each ten-year cycle the hares are either abundant or scarce, with relatively small changes in numbers. They may be reported as increasing and abundant or decreasing but still abundant. Similarly there may be so few in the locality at certain times that observers will not know whether they have increased or decreased, but they do know that they are scarce. Abundance at one place will not be the same absolute number of hares per square mile as at some other places, but the local residents know when hares are near the top part or the low part of their cycle of abundance for that locality. A curve of reports of increase tells the time of most widespread increase; a curve of reports of decrease tells the time of most widespread decrease; between them is the time of most widespread abundance. This latter, the time of peak abundance, is shown perfectly well by the curve of reports of a condition of abundance.

For the provinces of Canada, except Ontario, a compilation has been made in a less detailed way because the data for some provinces were too scanty to warrant the use of yearly maps. The areas covered by the reports were ignored, except to the extent of counting two votes for the larger areas and only one for the smaller areas. The number of votes for "abundance", expressed as percentages, was plotted as the curves of figure 11, for the provinces indicated in the figure.

The numbers of observers reporting from the different provinces and districts are given in table 1, and furnish some idea of the strength of the basis on which are founded the conclusions composing the next part of this paper. Unfortunately, many of these people did not report every year.

TABLE 1.—Number of observers.

Alaska	7
Alberta	37
British Columbia	52
Manitoba	43
New Brunswick	9
Newfoundland	8
North-west Territories	6
Nova Scotia	10
Ontario	555
Quebec	42
Saskatchewan	51
Yukon	4
Total	824

History of Fluctuations as Revealed by Fur Returns, Literature, and Questionnaires

The history of the abundance of the snowshoe hare in the Hudson bay watershed is depicted in the form of a graph in figure 3. For the years after 1903 when the Hudson's Bay Company figures end, the abundance in northern Ontario, as determined from questionnaire replies and reports in the literature, is shown by a second curve on the same graph. This latter area is largely included within the area of the fur returns and is the most nearly comparable for which data were available. This curve conforms well with the cycle shown by the Company data. The periodic nature of the cycles of abundance is clearly evident. From an examination of this record it is found that the peaks of abundance, or more particularly, the last year of great abundance before each great decrease or "crash", were the years: 1856, 1864, 1875, 1886, 1895, 1904, 1914, 1924, and 1934.

The cycles varied in length from eight to eleven years and the average length of the eight cycles was 9.6 years. In regard to the length of the cycle, it is shown, in figures 10 and 11 for the provinces and for sections of Ontario, that the cycle was

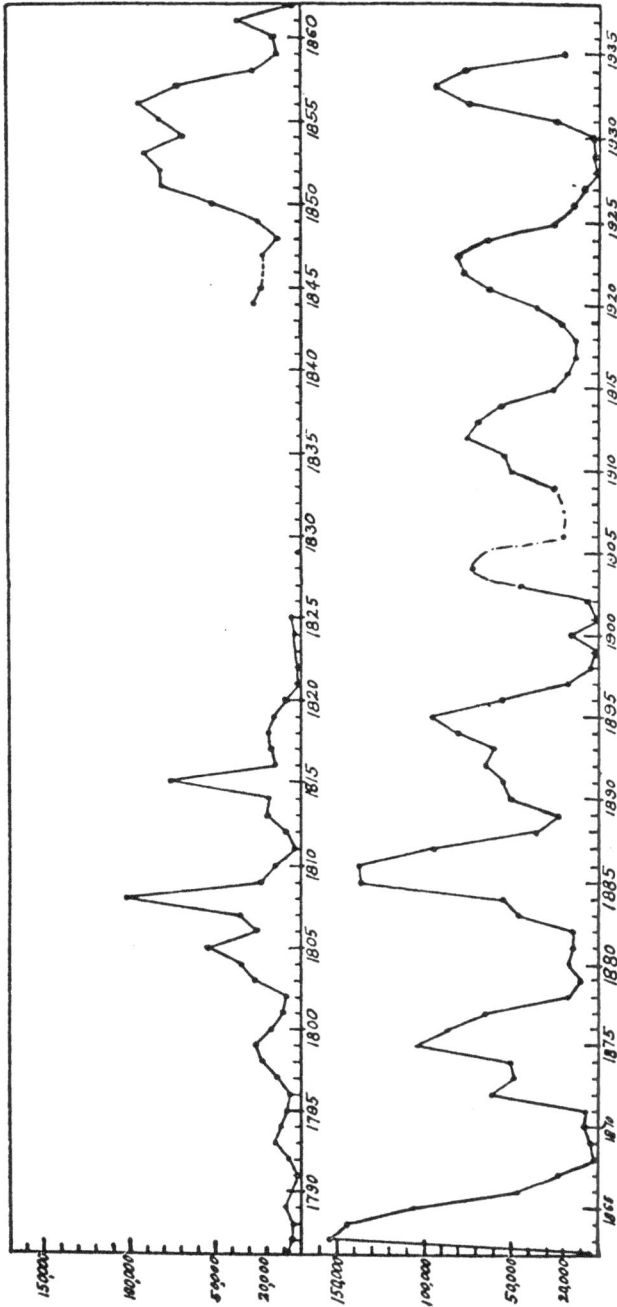

FIGURE 3.—History of abundance of varying hare in Hudson bay watershed. Up to 1903 based on Hudson's Bay Company fur returns, plotted over end of year of biological production; subsequent to 1903 compiled from replies to questionnaires.

about ten years long in every case for which the data extended far enough back.

The average number of rabbit skins taken by the Com-

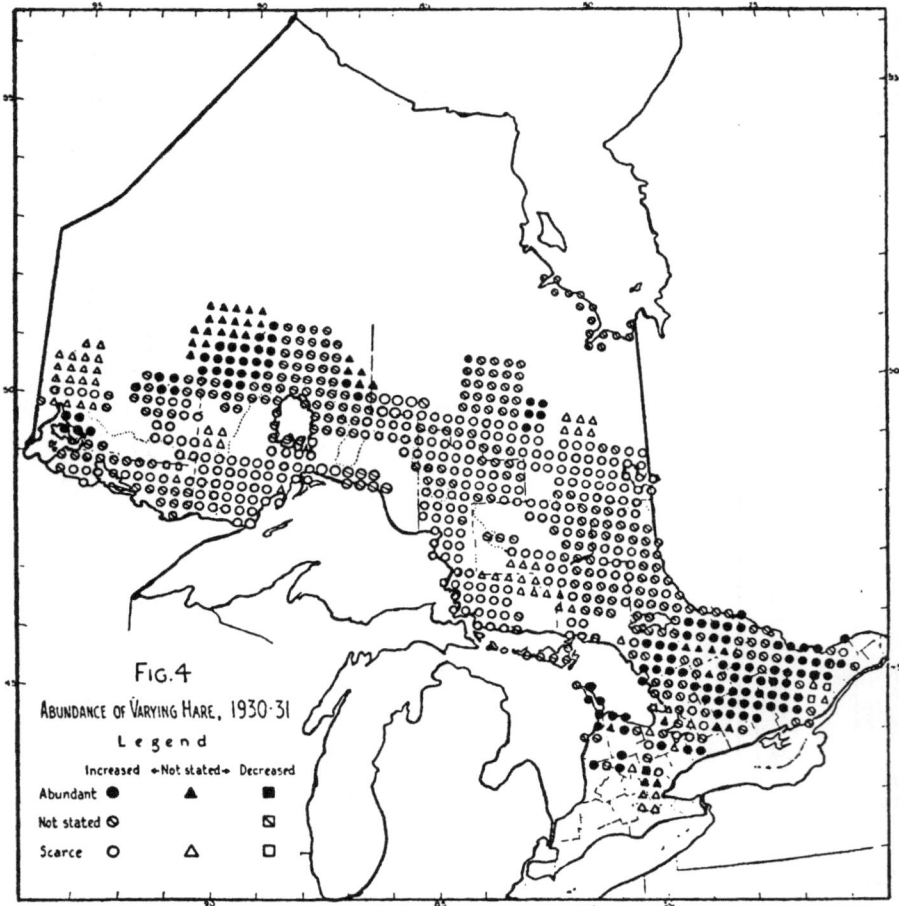

FIGURE 4

pany in the peak years of abundance was 115,000, and the average number of skins taken in the years of minimum abundance was 10,000.

Abundance in various regions of Ontario is shown in figure 10, based on the compilation of figure 2. Of course some localities will show deviations from this average picture. See the detailed yearly maps in figures 4-9. These graphs show that the last year of great abundance, which is the year in which

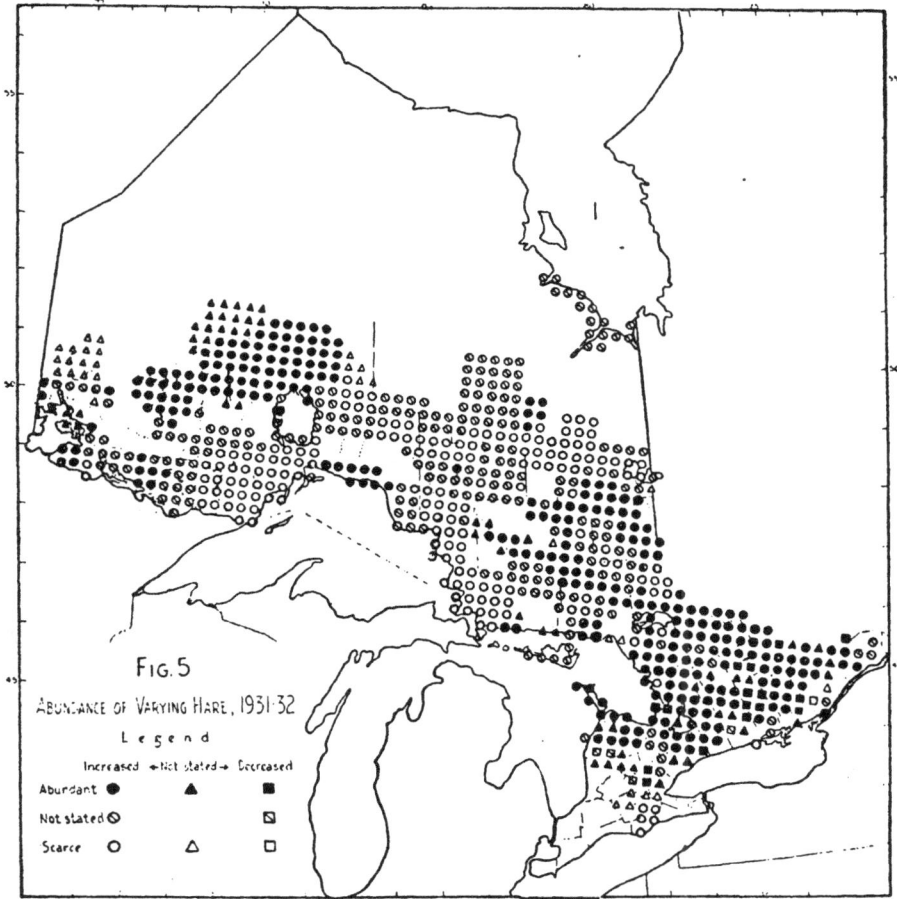

FIG. 5
ABUNDANCE OF VARYING HARE, 1931-32
L e g e n d
Increased ←Not stated→ Decreased
Aburdant ●   ▲   ■
Not stated ⊘   ⊠
Scarce   ○   △   □

FIGURE 5

the decrease began, was 1932 in a small district comprising chiefly the north parts of Frontenac county and the combined counties of Lennox and Addington; 1933 as general average for the peak year in southern Ontario, *i.e.*, from Bruce peninsula through Simcoe county, Muskoka district, and the counties of Haliburton and Peterborough and south of all these; the same year, 1933, in central Ontario from southern Algoma district, southern Sudbury, Nipissing, and Parry

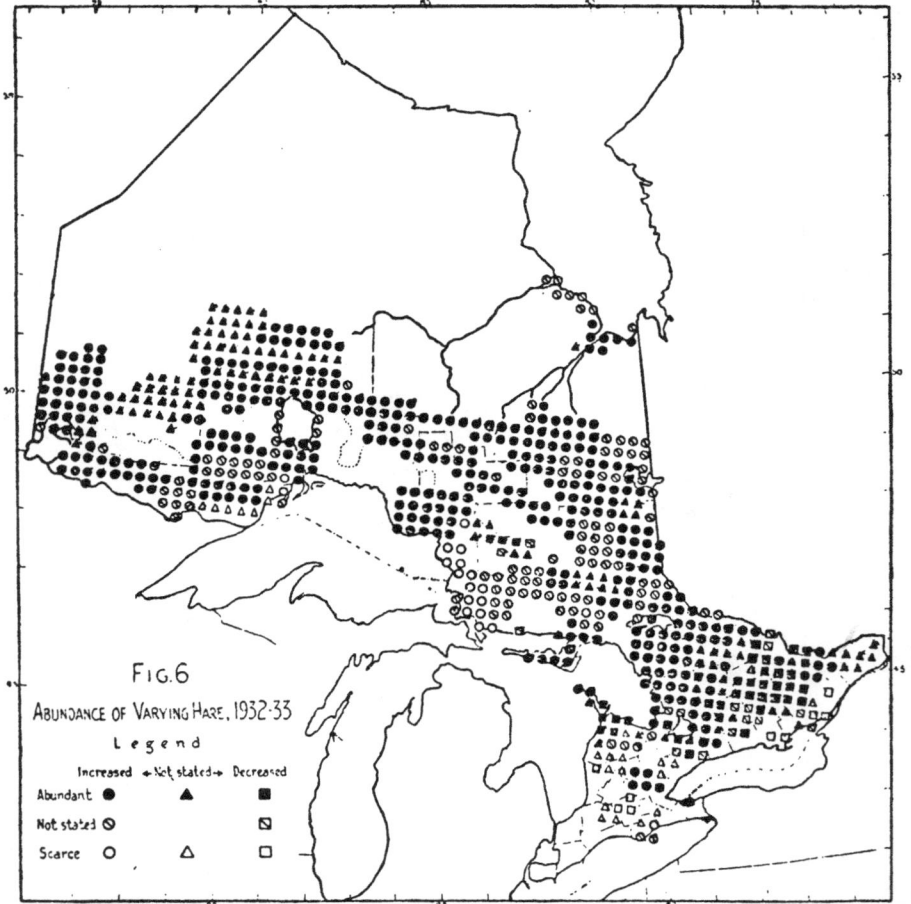

Fig. 6

Abundance of Varying Hare, 1932-33

Legend

Increased ← Not stated → Decreased

Abundant ●　　　▲　　　■

Not stated ◎　　　　　　 ◙

Scarce ○　　　△　　　□

Figure 6

Sound districts, and Algonquin Provincial Park to the northern part of Renfrew county; 1934 in the height of land country from Timiskaming district past lake Nipigon to include Kenora district; and 1935 on the northern part of the clay belt. That is, the cycle reached a peak first in south-eastern Ontario and the condition of maximum abundance moved slowly northward, reaching northern Ontario three years later.

The progression of the phase of the cycle across Ontario is

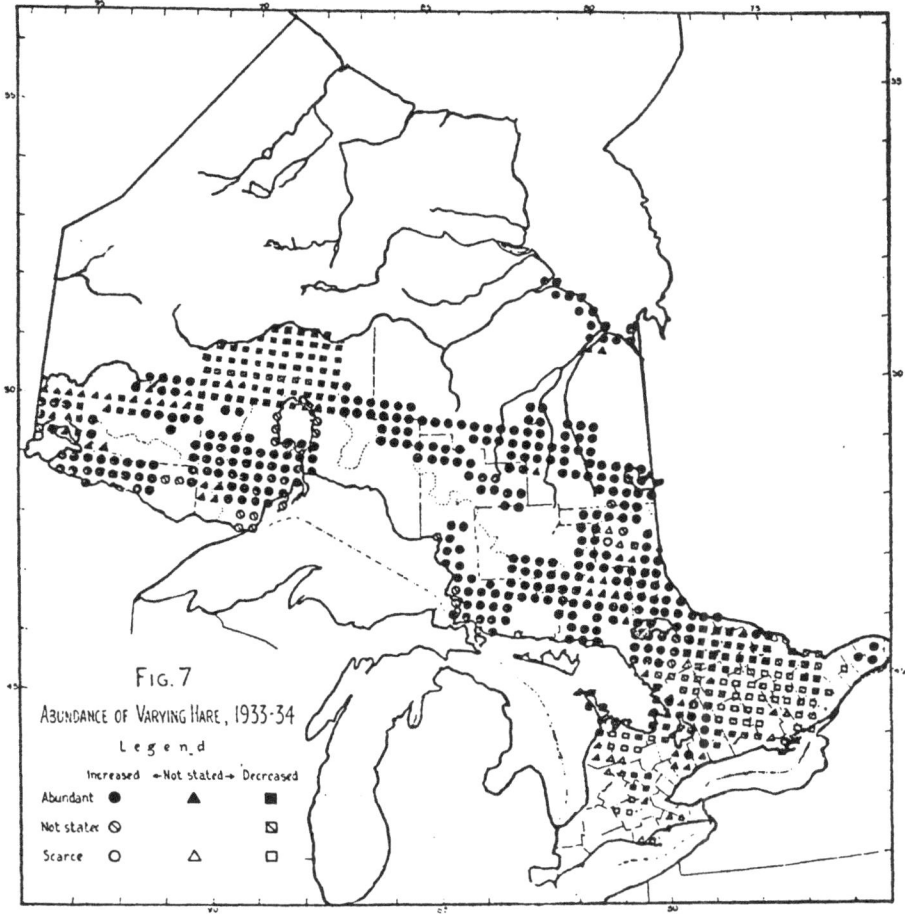

FIG. 7

ABUNDANCE OF VARYING HARE, 1933-34

Legend

Increased ←Not stated→ Decreased

Abundant   ●   ▲   ■

Not stated   ⊘   ⊠

Scarce   ○   △   □

FIGURE 7

illustrated in figure 12, by a map of the zones in which the peak of abundance occurred in the same year. The boundaries of the zones were determined by inspection of the maps of hare abundance for the years 1930-1 to 1935-6. These zones are true statements of the average conditions

during the recent peak of the hare cycle, and were true also
for the previous peak in the 1920's, at least.

Abundance in regions of Canada and in Newfoundland and

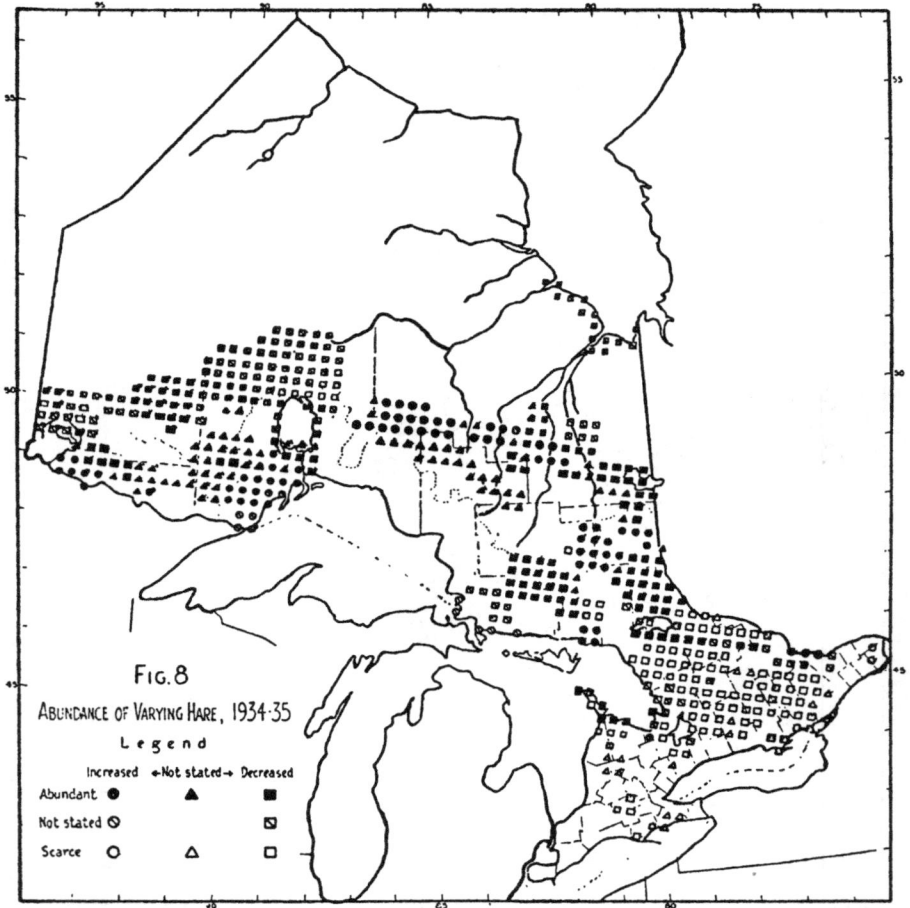

FIG. 8

ABUNDANCE OF VARYING HARE, 1934-35

Legend

Increased ←Not stated→ Decreased

Abundant ●   ▲   ■

Not stated ◙   ◪

Scarce   ○   △   □

FIGURE 8

Alaska, is shown in figure 11 by a curve for each region. It is
seen that the last years of great abundance were as follows:
(These statements have been arrived at by direct examination
of the reports and are not all shown in figure 11.) Newfoundland (insufficient data); maritime provinces and St. Lawrence

valley in 1932; central Quebec, 1933; southern parts of the prairie provinces, 1933; northern parts of the prairie provinces, 1934; Northwest Territories north to Great Bear lake

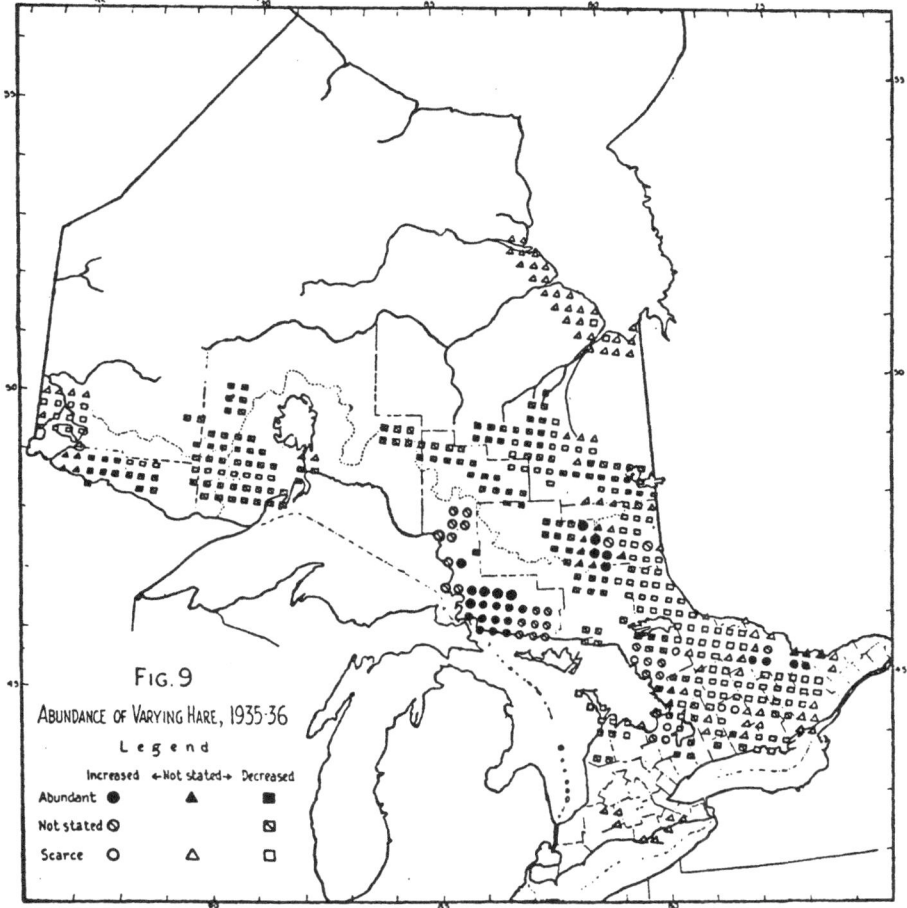

FIG. 9

ABUNDANCE OF VARYING HARE, 1935-36

Legend

Increased ←Not stated→ Decreased

	Increased	Not stated	Decreased
Abundant	●	▲	■
Not stated	⊘		◪
Scarce	○	△	□

FIGURE 9

(insufficient data but what there is puts the date as 1933 or more probably 1934); Mackenzie river delta, 1932 or 1933; Yukon and Alaska, 1935; southern British Columbia, 1932; northern British Columbia, 1933 and 1934. There appears to have been a progression through the country in the phase of

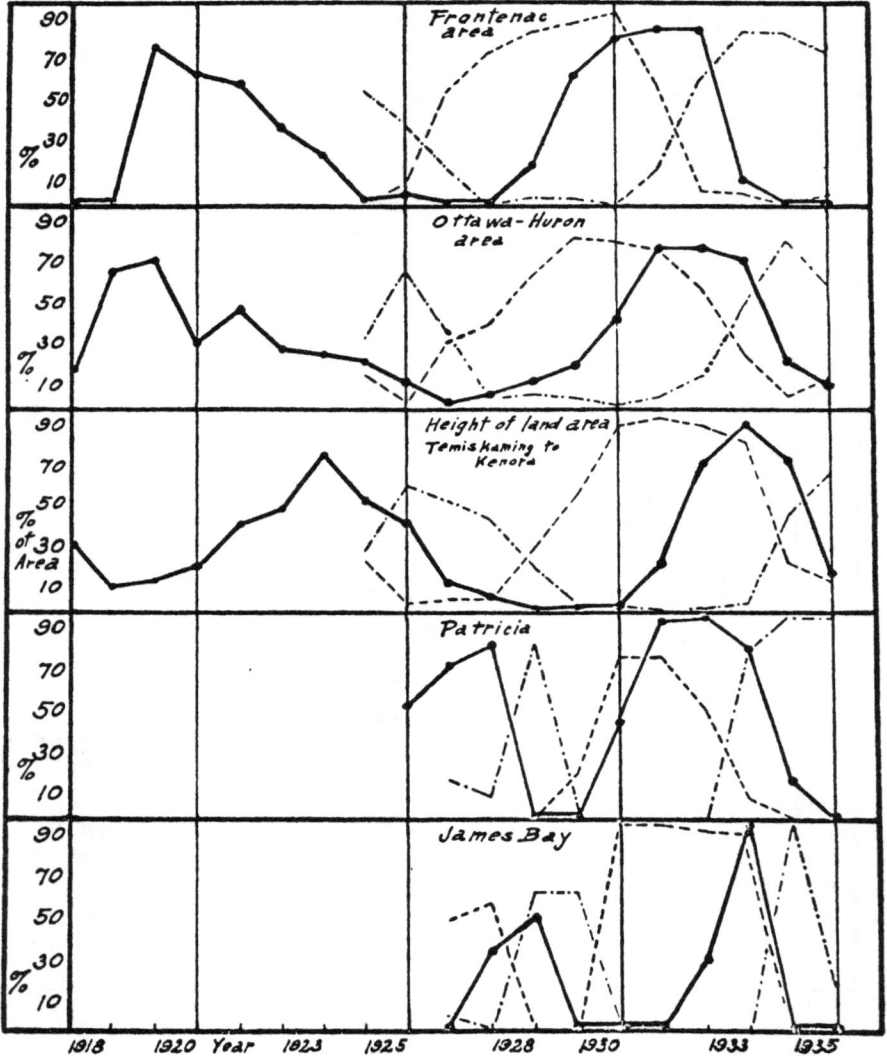

FIGURE 10.—Abundance of varying hares in different regions of Ontario, based on replies to questionnaires. Curves indicate the percentage of the area of each region covered by the replies, over which hares were abundant (heavy line), increased (line of dashes), and decreased (line of dots and dashes).

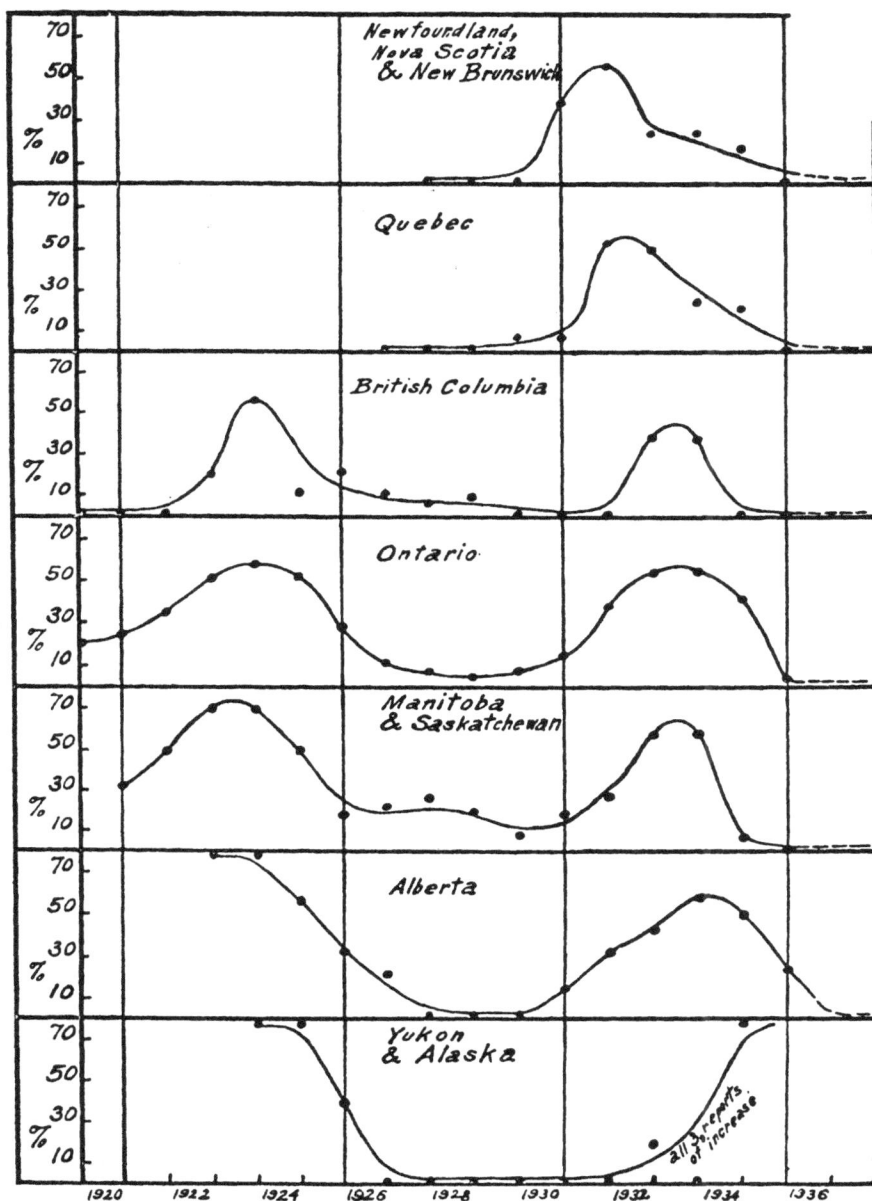

FIGURE 11.—Abundance of varying hares in different regions of Canada. Number of questionnaire replies reporting abundance plotted as percentage of total number of replies.

the cycle. The condition of abundance was reached and
passed first in the maritimes, south-eastern Ontario, southern
British Columbia, and the Mackenzie delta, and occurred last

FIGURE 12

in the interior of the continent in the northern part of the
Canadian life zone and the southern part of the Hudsonian life
zone, and in Alaska and Yukon.

Field Work

The localities, kinds of field data, and the methods of collecting and handling this material, will be first described; after which the results will be presented, and discussed. In this section on methods of field work, certain parts of the field routine, such as the collecting of specimens and care of live animals, will be dealt with although they are not directly concerned in the measurement of abundance.

Stations

Conditions were studied at first hand at the field stations listed in table 2. At the first four places much work was done, but the remaining places were visited for only a few hours each year. The field stations may be described as follows:

The four stations, Buckshot lake, Algonquin Park, Frank's bay, and Dorset, lie within the Ottawa-Huron forest region described by Sharpe and Brodie (1931). The soil has been

TABLE 2.—List of field stations and sample plots.

Station 1.—Buckshot lake (or Indian lake), work done in Miller and Clarendon townships of Frontenac county and Abinger township of Lennox and Addington county.

Plot 10, area 2½ acres, Abinger township, cedar and black ash swamp.

Plot 11, area 2½ acres, Abinger township, alder-willow swamp.

Plot 14, area 2½ acres, Clarendon township, cedar swamp.

Station 2.—Algonquin Provincial Park, Nipissing district, Ontario.

Plot 6, area 2½ acres, at Biggar lake in Biggar township, old forest of white pine, hemlock, balsam, white spruce, and white birch.

Plot 7, area 2½ acres, at Biggar lake, spruce-cedar swamp.

Plot 15, area 5 acres, at Biggar lake, same bush as plot 6, the area about the cabin.

Plot 26, area 2½ acres, at Cache lake in Canisbay township, white pine, hemlock, yellow birch.

Station 3.—Frank's bay, on lake Nipissing at the entrance to the French river, in Patterson township in Parry Sound district, Ontario.

Plot 2, area 2½ acres, larch-spruce swamp.

Plot 3, area 2½ acres, jack pine rock ridge.

Plot 4, area 2½ acres, second growth mixed white pine, poplar, and white birch.

Plot 16, area 5 acres, a part of it larch-spruce swamp, a part jack pine rock ridge, and the rest mixed white pine and sugar maple.

Plot 17, area 2½ acres, cedar swamp.

Station 4.—Smoky Falls, on the Mattagami river, in Harmon township, Cochrane
   district, Ontario.
   Plot 21, area 8½ acres, several vegetation types, largest area is jack pine,
      poplar, spruce.
   Plot 22, area 2½ acres, spruce-poplar forest.
   Plot 25, area 2½ acres, spruce-poplar forest.

Station 5.—Kettleby, King township, York county, Ontario.
   Plot 1, area 2½ acres, cedar swamp.

Station 6.—Bruce peninsula, Bruce county, Ontario.
   Plot 8, area 2½ acres, St. Edmonds township, brulé.
   Plot 9, area 2½ acres, Sauble beach, cedar swamp.

Station 7.—Minesing swamp, Simcoe county, Ontario.
   Plot 18, area 5/8 acre, near Mack, cedar swamp.

Station 8.—Balsam lake, Bexley township, Victoria county, Ontario.
   Plot 5, area 5/8 acre, cedar swamp.

Station 9.—Dorset, south to Wren lake, Sherborne township, Haliburton county,
   Ontario.
   Plot 19, area 2½ acres, sugar maple forest.
   Plot 20, area 2½ acres, mixed cedar and black ash swamp.

pictured by Dr. Coleman in the following terms: "The combination of kames (hills of sand and gravel with boulders)
with pure sand deposits, through which rise occasional hills of
the harder Archaean rocks, makes a region entirely unsuited
for agriculture and useful only for forest growth."

The vegetation may best be described by summarizing
Sharpe and Brodie's report by quotations. "Within the area
of higher elevations the tolerant hardwoods, hard maple and
yellow birch, predominate. Throughout the stand occur
scattered white pine and occasional clumps of hemlock. . . .
as the land rises from the swamps and water areas, pure coniferous stands are common, which in turn merge into mixed
stands of hardwoods and conifers. . . . On the uplands, where
the soils are shallow, the pineries exist, but where the soils are
deep, the mixed hardwoods and conifers still extend. . . . On
the deep, loamy slopes hardwoods predominate. Hard maple,
beech, basswood and yellow birch form the main stand, with
white ash and black cherry occurring individually. Inter-
mixed are white pine, hemlock and white spruce. . . . Farther

up the slopes on the drier sites the yellow birch, beech and basswood give way to a more scrubby growth of hard maple, ironwood and red oak, with an increase in the white pine content. . . . Extreme swamp sites in the Ottawa-Huron region are chiefly composed of cedar, spruce, tamarack and balsam in various proportions. Where the water circulates more freely and soil conditions are less acid, black ash, yellow birch, soft maple and elm occupy the site."

"The preceding is more a picture of the original forest, which as a result of lumbering and fire ravages, occupies only a limited area of the region. The cedar and spruce have been largely removed from the swamps and the white pine and hemlock from the mixed hardwood and pure coniferous stands. As a result of fire fully two-thirds of the region is covered now by a temporary type, the poplar-birch association. Such stands, for the most part, have replaced former pineries and occur on the thinner soils."

The main forest types around Buckshot lake were old hard maple forest, scrub red oak on rock ridges, alder-willow swamp, coniferous swamps, black ash swamp, and poplar-birch. The last covered the greatest part of the country.

There were farms scattered all through this district but in only a few places was the soil agriculturally productive. Many farms were abandoned, and on others the people eked out a poverty-stricken existence. Most of these people simply remained on the land after the pine logging was finished years ago.

The chief types of forest at Biggar lake in Algonquin Park were the hard maple bush on the high land, mixed white pine-hemlock-spruce-white birch on the slopes, and spruce swamp. The forest had been but slightly changed by man. The Booth Lumber Company had removed the finest of the large pines for square timber years before, but had not touched some less accessible area south of Biggar lake. The nearest human habitation was the Fassett Lumber Company depot, ten miles away, on Tea lake. An opportunity was presented to observe the hares and their fluctuations in numbers in so-called "virgin" forest. In 1935, Cache lake in the southern part of the park was visited and the work was done in old mixed tolerant hardwood-pine-hemlock forest from which the good pine trees had been culled.

The district about Frank's bay on lake Nipissing consisted of the poorer types, such as red oak, jack pine, poplar-birch, some second-growth mixed pine-hardwood, recent burns and swamps.

Some half dozen families of Indians lived there permanently; in summer a population of cottagers moved in, chiefly on the French river and not at Frank's bay itself; and in the autumn several hunting parties usually operated in the vicinity.

At the fourth station in the Ottawa-Huron region, namely, south of Dorset, the hard maple bush was dominant.

Smoky Falls marked the passing of the Mattagami river over the rock outcrops on the northern margin of the Precambrian shield. Leaving the flat clay belt country, the river dropped three hundred and sixty feet in about thirteen miles to enter the coastal plain of James bay. On the land the drop in elevation was not prominent and Archaean rocks were exposed only in scattered places. Young poplar-spruce forest covered much of the area. The mature forest was largely spruce with or without poplar. On the coastal plain this was confined to the immediate vicinity of the rivers. Seven miles north of Smoky Falls, that is, at the end of the rapids and beginning of the plain, the forest dwindled down to muskeg within a half mile distance of the river. The open muskeg began to appear as a major feature of the country less than a mile south of Smoky Falls. Jack pine occurred on some sand ridges in the transitional territory between the clay belt and the coastal plain.

The only inhabitants of the country were a few Indian families, mostly Crees, and at Smoky Falls, the staff of the power plant of the Spruce Falls Power and Paper Company.

In Bruce peninsula work was done at its base in a cedar swamp back of the dunes of Sauble beach, and near the tip of the peninsula five miles south of Tobermory in burnt-over country. The limestone rock had little or no soil on it at the latter place. Some patches of jack pine second growth, and a very little white pine were present. There was deeper soil with farms and mixed hardwood-conifer bush on the central and southern part of the peninsula.

The three stations, Kettleby, Minesing swamp, and Balsam lake were swamp and forest lands in farming country.

Methods

The work was done chiefly on sample study plots, which were marked to be readily found again, and were studied each year. These are listed under the proper field station in table 2. A map, usually a very simple sketch but in certain cases based on some survey work, was made of each plot. One purpose of the maps was to locate quadrats and snares at the same places each year. The usual size of the plots was five surveyors' chains square (*i.e.*, 110 yards on a side), with an area of two and a half acres. The size was modified to suit the available area and work to be done.

In field work four ways of measuring the abundance of hares were used: *e.g.*, trapping, census, observation, and scatology. The first two and the last were all done together on certain of the sample plots. This method has the objection that the removal of animals by trapping affects the population to be studied the next year. The only place where this factor seriously entered into the results is at Frank's bay in 1933. Three-quarters of the scatology and all the census and trapping were done that year on the area trapped heavily the previous year. It had not been restocked to the full density of population of the country in 1933 by immigration and survival of hares from the surrounding area. It is suggested that future work of this kind would be better done by using the vegetation types instead of arbitrarily bounded plots as the study units. In successive years do the trapping and scatology on new parts of the same types of vegetation. A map showing the areas of the various types would be an important adjunct to this system.

Trapping. The trapping was done chiefly by snares on the sample plots. The results were expressed as numbers of hares caught per hundred trap-nights; a trap-night being a trap set for one night, five trap-nights one trap left set for five nights or five traps for one night, and so on. This method was first suggested by Grinnell (1914). In the beginning of this work many kinds of snares were used, such as, spring poles, tip-up poles, and snubbed snares. The spring poles were made of a branch or young tree with the attached snare tied down by a slip knot to a log or other fixed attachment or held by a wooden trigger. These had three defects: first, a suitable spring tree was not often available at the place where a snare was wanted; secondly, the spring poles quickly lost their elasticity; and thirdly, too much work was involved. Tip-up poles were made by supporting a pole some five to eight feet long on a branch or large iron nail, or by binding with wire to a tree, so that one end hung just over a rabbit runway while the other longer and heavier part of the pole projected upwards beyond the support, and was propped up by a second small pole. When a rabbit entered the snare attached to the lower end of the pole, its pushing and jumping quickly would shake away the prop and the pole would tip up, hanging the animal. This form of snare was used considerably due to its reliability. The snubbed snares consisted simply of the snare wire tied round a small tree, or to a branch or shrub or log or to a nail driven into a tree for the purpose. They had the disadvantage of sometimes letting an animal live too long and hares occasionally broke away. However, if the animal made a strong effort to get away immediately it felt the wire pull against it, the noose would tighten so far as to kill the hare without delay. Few hares were lost after stopping the use of soft brass wire, the usual rabbit "snare wire", and using tinned iron wire, Brown & Sharpe's No. 22 brass wire gauge. The advantage of snubbed snares lay in their simplicity and speed of construction, coupled with the fact that they were as effective as the more elaborate types. A large noose was used, about seven to nine inches in diameter with the bottom slightly flattened. The bottom of the noose was set about three inches off the ground. The noose was always set with the end part of the wire forming the bottom of the noose, so that less wire would be pulled across the animal's chest or throat than if the noose were placed the other way up. Often the noose could be more or less concealed by vegetation or by sticking twigs in the ground. In Algonquin Park hares were caught alive several times in a wire noose on the end of an

eight-foot pole held in the hand as one watched them on a
dark night by means of a flashlight. Snares were well distributed over the study plots. Only a few irregular trap lines
were set away from plots. In cedar and spruce swamps there
are usually evident runways on which to get snares, but in
dried hardwood bush there is very little to guide one in
summer.

Several kinds of traps for catching hares alive were tried.
Live-traps have been described by Anderson (1932), Bailey
(1921, 1932, 1933, 1934), Burt (1927), Dice (1925), Hatt
(1925), Scheffer (1934), and Silver (1929). The type with
a door dropping in a vertical groove and released by a trigger
was not satisfactory. It was bulky and liable to get out of
order by the swelling or warping effects of wet weather or
by ice and snow filling the groove in winter. A better type
of trap, which was designed by Mr. V. Morrow of Smoky
Falls, consisted of a wooden box with an extra floor inside,
hinged at the back. The door opened out and down in
front, being hinged on a rod about an inch above its bottom
edge. The lower end of the door turning up into the box
supported the front end of the hinged floor. Once an animal
got inside with its hind feet off the door its weight pushed the
floor down, closing the door. The door was locked because
the floor dropped right down, blocking the bottom of the
door. A small hole was left in a lower corner at the floor
level to admit a finger to raise the floor when ready to take
the animal out of the trap. The outside dimensions of the
traps were $9 \times 10 \times 13$ inches and they accommodated hares
and even skunks. The trap shown in figure 13 was the first
experimental one and twice as large as those made later.
These traps, although unpainted, operated successfully in all
weather from winter through spring to summer. Various
baits, such as dried fruits, bread, oats, salt, carrots, and apple
were used with fair success; the fruit was the most generally
successful bait. Even without bait a hare entered a trap
on several occasions, seemingly interested in the dark hole.

Censuses. Actual numerical censuses were obtained in
various ways. At Frank's bay the block of bush included
in the three plots numbers 2, 3, and 4, a total of about eight

acres without definite natural boundaries, was trapped to
near extinction yearly.   Any hares observed on the area after
the trapping were added to the total caught, giving a population figure of a rough sort.   The area was a mixture of kinds
of forests typical of the country.   These censuses necessarily
told much the same story as the trapping and were not
independent measurements.

FIGURE 13.—Hare den and trap.

At Biggar lake in Algonquin Park the hares frequenting
the vicinity of the cabin were distinguished on sight and
counted.   This was aided by catching two in 1932 and two
more in 1933, putting numbered aluminum tags on their ears
in different positions, and releasing them again, as suggested
by Dice (1931).   The area concerned was unknown, but from
some observations of movements of individual hares and the
lay of the land a guess of five acres was made.
    At Smoky Falls on the Mattagami river the census was
obtained by trapping hares off plot 21 until there were

apparently few if any left on the area. Plot 21 was a natural unit, nearly surrounded by barriers to the movement of hares, and was typical of the forested part of the country. Only one hare track was found crossing the railway which formed one side of the plot, but the snow was not in condition all the time to record the passing of a hare. This was a better census than any obtained elsewhere because the true area concerned was known more surely.

Observations. The distances travelled by the observer and the hares seen at each field station were recorded and expressed as the numbers of hares observed per fifty miles of travelling. This general idea was suggested by Taylor (1930). The method was not refined by separating distances covered in daylight from those in the night because not enough walking was done in the dark at Frank's bay, Algonquin Park, or Smoky Falls. The night-time figures for Buckshot lake district are included in table 5, but no further use is made of them at present. The much greater distances covered and resulting greater accuracy in the figures for Buckshot lake than for other places were due to the use of a car. The narrow roads through the woods had not enough car traffic to become barriers and be avoided by hares, but were freely frequented by them. The population of hares should vary with the square of the number observed per fifty miles of travelling, because that figure is a linear measure whereas population concerns the whole area. This measure can be calibrated by the census of hares on plot 21 at Smoky Falls, where the density of population was of the order of one thousand hares per square mile, and the number observed per fifty miles of walking was six. The areas of open muskeg in that country did not enter into either the census plot or the distance walked and so the two remain comparable. By this calibration a figure of twelve hares per fifty miles of travelling should indicate a population of four thousand hares per square mile, and a figure of three hares per fifty miles should indicate a population of two hundred and fifty per square mile. Of course these estimates are rough and subject to much error, but, in the lack of anything better, they appear to have some value as coarse approximations.

Scatology. The abundance of droppings (scats or pellets) obviously is related to the abundance of hares. Scatology figures have been used as measures of abundance of jack rabbits, *Lepus alleni* and *Lepus californicus* by Vorhies and Taylor (1933), and the idea was suggested by Taylor (1930), and by Ruhl (1932). Their statement quoted below is true for the varying hare: "In captivity rabbits regularly defecate while they are feeding or very soon after. Field observations indicate similar habits on the part of hares in the wild." In other words, hares defecate while they are actively moving about the bush. Elton (1927) tells that rabbits, *Oryctolagus cuniculus*, "have the peculiar habit of depositing dung in particular spots", but this is not the case with the varying hare. Dice (1931) made a study of the abundance of the varying hare in northern Michigan by noting the occurrence of droppings in likely places, and found them on 89 per cent. of the places. In the present work the average number of droppings per square metre was arrived at for each plot by counting the droppings on, generally, twenty-five quadrats each 0.2 square metre in area, and dividing the total by 5. In the first season of field work before the full value of the method was evident, fewer than twenty-five quadrats were usually taken on a plot. A refinement that was found practicable consisted in classing droppings as "old" or "recent". "Old" ones were those obviously weathered, or under debris in the litter or humus or with moss or other plants grown over them. "Recent" droppings were all the rest. The old ones represented the abundance of hares at some time previous to that represented by the "recent" count. The count of "old" ones in September would refer largely to conditions of early summer. One season's droppings would be mostly disintegrated and buried before the next summer.

The per cent. frequency of occurrence of droppings on the quadrats has been calculated but it was decided not to use it, as the number per square metre seemed a more meaningful concept and was besides independent of the size of the quadrats. On plot 21 there occur two quadrats with nineteen and twenty "recent" droppings respectively. These numbers are over twice as high as any others from the plot and would have an excessive influence on the average value; so they have been omitted from all calculations. Similarly a quadrat on plot 8 in 1934 had over six times as many "old" droppings as the next highest count and it has been omitted from compilations.

It was attempted to gain an idea of the accuracy of the averages that have been calculated from sets of twenty-five quadrats. One would expect that the distribution of droppings over the ground would not be quite random, because logs and holes and preferences on the part of the hares for particular places and food plants will all operate to influence the distribution. To a large extent the hares wander over all parts of an area as is very evident by the tracks on snow in winter, but certain places and runways are used more than other places. It is, however, conceivable that, considering a sample of $25 \times 0.2 = 5$ square metres, the small differences might average out and the resulting frequency distribution be random or nearly so. In that case the distribution will be of the binomial form: $(q+p)^N$ where $p =$ the fraction of the country comprised in the area of the quadrats and $q = 1 - p$, and $N =$ total number of droppings in the country. The most probable number of droppings on the quadrat area will be Np, and the average will approach this value more or less closely. However, since p is a very small fraction, the distribution will be well represented by the Poisson form: $\epsilon^{-m}\left(1 + \dfrac{m}{1} + \dfrac{m^2}{1.2} + \dfrac{m^3}{1.2.3} + \ldots\right)$ where $m = Np$ (Fisher, 1930).

To test whether or not an observed distribution is Poisson, two procedures are available:

(a) Calculate the variance $=$ the square of the standard deviation. If the distribution is Poisson, it should equal the *mean*. If it is not nearly equal, use χ^2 test to see if it differs significantly; $\chi^2 = \sum \dfrac{(X - \bar{x})^2}{\bar{x}}$, where $\bar{x}$ is the mean.

(b) Calculate the terms of the Poisson distribution having the observed mean (m) value, and compare with the corresponding observed terms (end terms should be grouped, so that no frequency class, observed or expected, contains less

than seven or eight terms).   It is considered that the test (a)
is better and more suitable to this problem.

A set of a hundred quadrats, taken in May, 1934, on plot 3,
when hares were moderately scarce, was divided into four
groups randomly, without replacing ones previously chosen.
The total numbers of droppings to these four groups were
20, 9, 11, 5, respectively, and their average 11.2.   The
variance (= square of standard deviation) of these four
samples was 40.3 which is much greater than the mean.
Using the χ^2 test of significance, the result is 10.8 which is
just below the 0.01 level of significance for 3 degrees of free-
dom.   Hence the data do not prove that distribution is not
random in the twenty-five quadrat samples, but they suggest
it.   Hence the fiducial limits (Clopper and Pearson, 1934;
Ricker, 1937a, b) will afford a minimal (since we are not sure
of the distribution) estimate of range of abundance; that is,
they will likely be too narrow.   The confidence or fiducial
limits, for a confidence coefficient of 0.99 or a probability of
0.01 as a limit, are 4.3 and 22.6; that is, the true average total
number of droppings on twenty-five quadrats or five square
metres may be between 4 and 23.

Similarly the mean of three groups of twenty-five quadrats
each, which comprised a set of seventy-five quadrats, taken
on plot 21, 1935, when hares were abundant, was 50.   It is
desired to compare this result with the result on plot 3 when
hares were scarce, as an example, to see whether the droppings
counts can be demonstrated to be significantly different.
The fiducial limits for this average are 33 and 71.   This range
does not nearly meet, let alone overlap, the range for the
other data, so the scatology counts do significantly differ-
entiate between an abundance of hares and a moderate
scarcity of hares.

As was done with the hares observed, so the scatology
may be roughly calibrated by the census on plot 21.   Because
the droppings have been expressed on an area basis, they
should vary directly with the population.   The density of
population was about one thousand hares per square mile
and the number of droppings per square metre was nine and
a half, on plot 21.   By this calibration a droppings count of
thirty-eight would indicate a population of four thousand

hares per square mile; and an average of two and three-quarters droppings would correspond to three hundred hares per square mile.

Autopsies of hares. In the field work, besides measuring the abundance of hares, specimens were examined for parasites and diseases with the hope of finding the direct causes of the decreases in abundance. (See the reports on external and internal parasites and bacteriology for the methods used at autopsy.) In addition to the hares caught in snares and box traps, some were collected by shooting with a .22 rifle. The rifle was used rather than a shotgun because it made only one hole in the specimen instead of many, and because the bullet could usually be put through the upper part of the shoulder or some other place where it would not interfere with a good autopsy. Hares were most often shot in the evening when they became active. An aid to night shooting was an electric flashlight, the lamp of which could be strapped on one's forehead. The beam of light could be directed along the gun barrel, illuminating the sights and shining on the target. The hares used in this investigation totalled two hundred and seventy, of which about one hundred and five only were available for autopsy.

Experimental animals. Varying hares, domestic rabbits, and guinea pigs were kept alive in cages in camp; the guinea pigs especially were available at all the main field stations. The cage in which the pigs were kept was made with a wire mesh floor, wire on part of the front and top, and a trap-door lid on top. When travelling a galvanized iron tray under the cage caught the droppings. The rabbits and guinea pigs were used in the bacteriological work. Some life history studies were obtained from caged varying hares. They were sometimes used in experimental studies of parasites and diseases. Hares that were snared or shot could give little information about causes of mortality among hares in the woods. To circumvent this difficulty hares were caught alive at Smoky Falls in 1935, and the traps have been described above. The cages used for the hares were similar to the guinea pig cage, except that each cage had a removable partition in the middle and two trap-doors, so that two animals could be kept in each cage when many animals were on hand.   At Smoky Falls the hares were given runs, five feet long by one foot wide, made entirely of one inch mesh chicken wire.   Seven of them were built together on a wooden frame.   The wire floor was a foot off the ground to prevent infection of the animals, and to facilitate cleaning.   Each hare cage could be placed across the doors of two adjacent runs after removing the two window-like sections of wire mesh on the front of the cage.   A door hinged at its top was hung in the entrance to each run, so that the animal could be confined to either the cage or the run on occasion.   Several hares caught their heels in the wire.   It should have been half-inch mesh, but that was not obtainable.

As it was desired to keep some of the hares free from any infection that might possibly be carried by black flies or mosquitoes, a fly-proof tent was constructed.   It was made first of mosquito netting, but the flies and mosquitoes readily went through this.   A complete covering of cheese cloth was therefore sewn on.   The size of the tent was nine by twelve feet with a wall three and a half feet high.   In order to enter and leave the tent without admitting flies, the front part was partitioned off by a "fly-bar" one yard in from the front "fly-bar" of the tent.   The "fly-bar" was simply a partition or wall of cheese cloth and netting, sewn in all round the top and sides, and made so much wider than the tent that one could raise it waist high and go under.   The bottoms of the "fly-bars" were protected and made heavy by sewing sacking to them, and were weighted down by laying heavy poles on them.   One entered by going under the front "fly-bar" into the small ante-room, killing all the flies and mosquitoes that had come with one by means of the pump gun of fly spray which was kept always there, and then going under the inner "fly-bar" to where the animals lived.   The fly spray consisted simply of ordinary coal oil with pyrethrum powder, one-quarter pound to the gallon, mixed in and allowed to settle after shaking thoroughly several times.   The set of runways described above was in this animal tent. During the two months of its use only one black fly and two mosquitoes were found inside the animal room of the tent.

Both hares and guinea pigs were fed on oats, freshly

pulled or cut grass, clover and other plants, poplar and willow twigs, occasionally cabbage or carrots, and water.

With these methods the hares were thriving, and most of them became quite used to my presence and occasional handling of them.

Results

The results at field stations as shown by the four ways of estimating abundance are tabulated in tables 3-6 and are shown in figure 14. They will be discussed below for each locality and then compared with the findings from the questionnaires.

(*a*) At Buckshot lake the number of hares caught per hundred trap-nights decreased from four in September of 1932 to one in 1933 and none in 1934 and 1935. The number of hares seen per fifty miles of travelling was eleven in July, 1932, but decreased to three by September of that year, in spite of the natural production of young hares that one would expect to have taken place. For the next three years very few hares were observed. "Recent" hare droppings were abundant in July, but less so in September. There were three times as many "old" ones in September as in July. Droppings were scarce during the next three years. Local residents claimed hares had been actually more numerous in the spring than in July, 1932. The three ways of measuring the abundance of hares show conclusively, then, that hares were abundant in the early part of 1932, but decreased to such an extent that they were almost scarce before the end of the year, and that there have been hardly any in the country since that time.

The estimates of the actual densities of population that may be made by the methods described earlier are of interest. As the droppings were counted only in swamps at Buckshot lake, the populations estimated from them represent the densities of population in the swamp areas alone, not over the whole country. In the first line of table 7 these populations in the swamps are shown, and the second line of the table contains the populations of the whole country calculated from the numbers of hares observed. The light thrown on

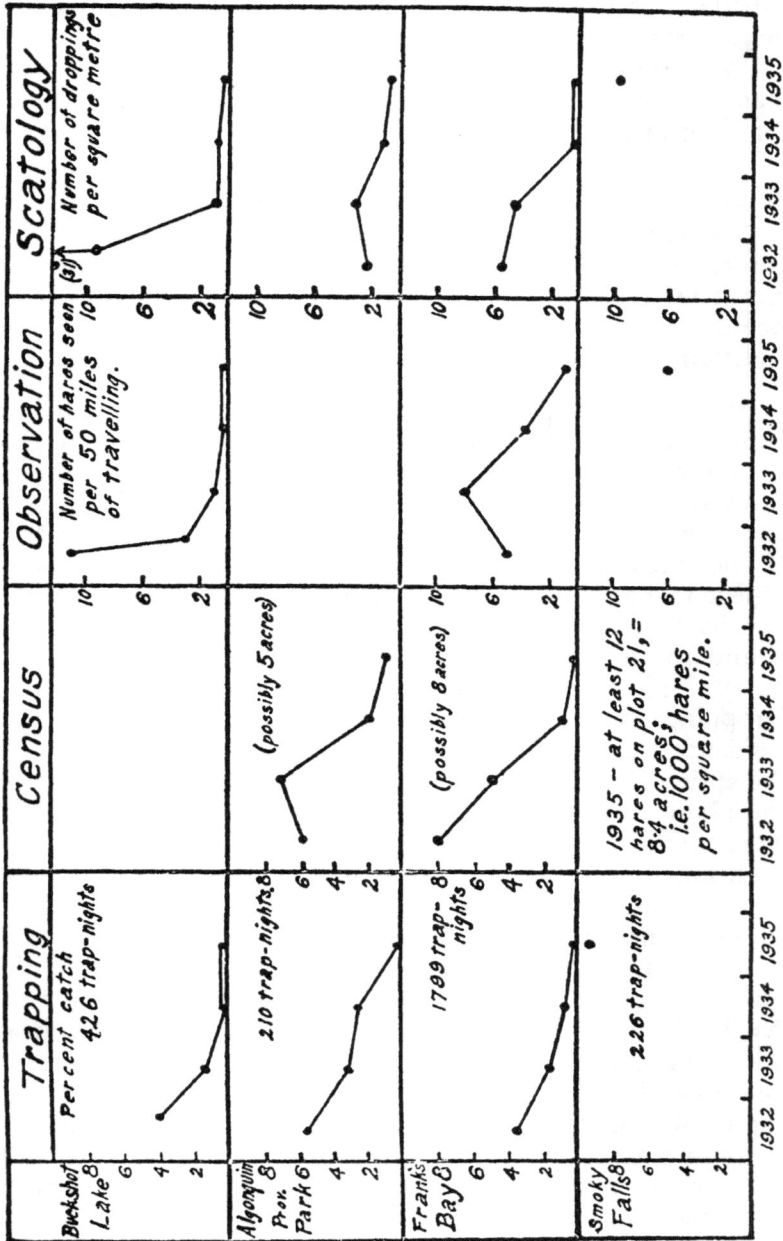

FIGURE 14.—Abundance of varying hares as shown by results of field work.

TABLE 3.—Summary of trapping (by snares except where box traps specified).

Plot number	Numbers of trap-nights				Numbers of hares caught				Per cent. catch			
	1932	1933	1934	1935	1932	1933	1934	1935	1932	1933	1934	1935
At Buckshot lake												
10	16	36	38	50	1	1	0	0				
11	8	12	24	24	0	0	0	0				
14		31	102	85		0	0	0				
Totals ..	24	79	164	159	1	1	0	0	4	1	0	0
In Algonquin Provincial Park												
7	12	45	56	14	1	1	2	0				
15		19				1						
6	6	6	24	8	0	0	0	0				
Box traps				20				0				
Totals ..	18	70	80	42	1	2	2	0	6	3	2	0
Frank's bay												
2	93	147	123	133	8	6	1	0				
3	93	120	90	98	0	0	0	0				
4	121	107	134	133	1	0	0	0				
16 hdwd.			121	42			1	0				
16 j. pine			120	42			1	0				
16 bog			124	42			1	0				
Box traps				16				0				
Totals ..	307	374	712	406	9	6	4	0	2.9	1.6	0.6	0
Smoky Falls												
21 box traps				103				11				
trap line		snares	21					3				
trap line		box traps	102					7				
Totals ..			226					21				9

the habitat relations of the varying hare by the above figures will be elaborated and discussed in the note on matters pertaining to the life history of the hare.

The magnitude of the fluctuations shown above is astonish-

TABLE 4.—Summary of censuses.

Station	Area	Number of hares 1932 1933 1934 1935			
Algonquin Park	about 5 acres	6	7	1	1
Frank's bay	about 8 acres	8	6	1	0
Smoky Falls	about 8½ acres				12

TABLE 5.—Summary of hares observed.

	Distance travelled					Hares observed					Number of hares per fifty miles				
	1932 July	Sept.	1933	1934	1935	1932 July	Sept.	1933	1934	1935	1932 July	Sept.	1933	1934	1935
Buckshot lake															
Day	125	64	727	407	245	18	0	2	0	0	7	0	0.2	0	0
Night......	41	81	288	170	50	17	9	17	1	0	21	5½	3	0.3	0
Totals.	166	145	1015	577	295	35	9	19	1	0					
Over-all averages											11	3	1	0.1	0
Averages of day and night figures											14	3	1½	0.15	0

	Distance travelled				Hares observed				Number of hares per fifty miles			
	1932	1933	1934	1935	1932	1933	1934	1935	1932	1933	1934	1935
Frank's bay	45	66	165	23	5	9	10	1	5	8	3	1
Smoky Falls				200				24				6

ing, ranging from near zero to over three thousand. The field work has not covered the part of the cycle in which the hares increase in numbers so it is not yet worth while to construct graphs of the population changes.

(b) In Algonquin Park the trapping figure was high in 1932 but did not become really low until 1935. In 1932 there were at least six hares in the vicinity of the cabin, that is on the census area, and at least seven in 1933, but only one in 1934 and 1935. The amount of walking done in the park was insufficient to serve for calculating the numbers of hares observed per fifty miles of travelling. The scatology

TABLE 6a.—Summary of scatology ("old" droppings only).

Buckshot lake

Plot number	Numbers of quadrats 1932 July	1932 Sept	1933	1934	1935	Numbers of droppings 1932 July	1932 Sept	1933	1934	1935	Frequency of occurrence 1932 July	1932 Sept	1933	1934	1935
10	10	10	10	25	25	39	34	14	5	1	9	8	6	5	1
11	10	10	10	25	25	71	158	93	33	10	7	10	10	13	5
14	..	..	25	25	25	..	..	24	9	2	..	..	13	4	2
Totals..	20	20	45	75	75	110	192	131	47	13	16	18	29	22	8

	Areas in square metres					Numbers per square metre					Per cent. frequencies				
	4	4	9	15	15	27	63	15	3	1	80	90	65	29	11

Algonquin Park

Plot number	1932	1933	1934	1935	1932	1933	1934	1935	1932	1933	1934	1935
6	15	25	25	25	1	6	2	2	1	6	2	2
7	10	25	25	25	10	12	4	1	5	7	3	1
26				25				1				1
Totals..	25	50	50	75	11	18	6	4	6	13	5	4

	Areas in square metres				Numbers per square metre				Per cent. frequencies			
	5	10	10	15	2.2	1.8	0.6	0.3	24	26	10	5

Frank's bay

Plot number	1932	1933	1934	1935	1932	1933	1934	1935	1932	1933	1934	1935
2	11	11	25	25	0	14	5	0	0	6	5	0
3	25	25	100	25	7	63	9	1	5	15	8	1
4	15	25	25	25	22	55	5	6	4	19	5	5
16 hdwd.	}		25	25	}		22	8	}		11	6
16 j. pine	}	25	25	25	}	57	10	8	}	15	4	4
16 bog	}		25	25	}		2	3	}		2	3
17			25	25			29	8			11	6
Totals..	51	86	250	175	29	189	82	34	9	55	46	25

	Areas in square metres				Numbers per square metre				Per cent. frequencies			
	10.2	17.2	50	35	2.8	11	1.6	1	18	64	18	14

TABLE 6a.—*Continued*

Plot number	Numbers of quadrats				Numbers of droppings				Frequency of occurrence			
	1932	1933	1934	1935	1932	1933	1934	1935	1932	1933	1934	1935
Smoky Falls												
21				73				155				48
22				20				23				8
25				25				12				9
Totals..				118				190				65

	Areas in square metres		Numbers per square metre		Per cent. frequencies	
		23.6		8.5		55

Plot number	Numbers of quadrats				Numbers of droppings				Frequency of occurrence			
	1932	1933	1934	1935	1932	1933	1934	1935	1932	1933	1934	1935
Bruce peninsula												
8	10	25	24	25	31	30	10	26	5	11	6	10
9	10	20	25	25	16	56	33	25	7	18	12	11
Totals..	20	45	49	50	47	86	43	51	12	29	18	21

	Areas in square metres				Numbers per square metre				Per cent. frequencies			
	4	9	9.8	10	12	9.5	4.4	5.1	60	65	37	42

Plot number	Numbers of quadrats				Numbers of droppings				Frequency of occurrence			
Minesing swamp												
18			25	25			11	.7			8	4

	Areas in square metres		Numbers per square metre		Per cent. frequencies	
	5	5	2.2	1.4	32	16

Plot number	Numbers of quadrats				Numbers of droppings				Frequency of occurrence			
Balsam lake												
5	16		25	25	5		13	16	3		10	8

	Areas in square metres			Numbers per square metre			Per cent. frequencies		
	3.2	5	5	1.6	2.6	3.2	19	40	32

Plot number	Numbers of quadrats				Numbers of droppings				Frequency of occurrence			
Dorset												
19			25	25			0	0			0	0
20			25	25			3	0			3	0
Totals..			50	50			3	0			3	0

	Areas in square metres		Numbers per square metre		Per cent. frequencies	
	10	10	0.3	0	6	0

TABLE 6b.—Summary of scatology ("recent" droppings only).

Buckshot lake

Plot number	Numbers of quadrats — July 1932	Sept 1932	1933	1934	1935	Numbers of droppings — July 1932	Sept 1932	1933	1934	1935	Frequency of occurrence — July 1932	Sept 1932	1933	1934	1935
10	10	10	10	25	25	44	12	2	2	0	7	5	2	2	0
11	10	10	10	25	25	79	24	2	3	1	9	7	2	3	1
14	..	25	25	25	25			2	7	3		20	2	5	3
Totals..	20	20	45	75	75	123	36	6	12	4	16	32	6	10	4
Areas in square metres / Numbers per square metre / Per cent. frequencies	4	4	9	15	15	31	9	0.7	0.8	0.3	80	71	13	13	5

Algonquin Park

Plot number	Numbers of quadrats — 1932	1933	1934	1935	Numbers of droppings — 1932	1933	1934	1935	Frequency of occurrence — 1932	1933	1934	1935
6	15	25	25	25	5	13	4	2	4	7	3	1
7	10	25	25	25	6	17	2	1	3	8	2	1
26				25				1				1
Totals..	25	50	50	75	11	30	6	4	7	15	5	3
Areas in square metres / Numbers per square metre / Per cent. frequencies	5	10	10	15	2.2	3	0.6	0.3	28	30	10	4

Frank's bay

Plot number	Numbers of quadrats — 1932	1933	1934	1935	Numbers of droppings — 1932	1933	1934	1935	Frequency of occurrence — 1932	1933	1934	1935
2	11	11	25	25	3	8	0	2	3	6	0	2
3	25	25	100	25	36	23	15	1	10	8	8	1
4	15	25	25	25	15	10	2	12	9	9	2	6
16 hdwd.			25	25			2	2			2	1
16 j. pine		25	25	25		34	4	3		10	3	1
16 bog			25	25			6	0			2	0
17			25	25			8	17			6	6
Totals..	51	86	250	175	54	75	37	37	22	33	23	17
Areas in square metres / Numbers per square metre / Per cent. frequencies	10.2	17.2	50	35	5.3	4.3	0.7	1.1	43	38	9	10

TABLE 6b.—*Continued*

Plot number	Numbers of quadrats 1932	1933	1934	1935	Numbers of droppings 1932	1933	1934	1935	Frequency of occurrence 1932	1933	1934	1935
Smoky Falls												
21				73				138				51
22				20				31				11
25				25				52				18
Totals..				118				221				80
Areas in square metres				23.6	Numbers per square metre			9.4	Per cent. frequencies			68
Bruce peninsula												
8	10	25	24	25	11	62	2	5	7	17	2	5
9	10	20	25	25	25	44	34	18	6	15	14	10
Totals..	20	45	49	50	36	106	36	23	13	32	16	15
Areas in square metres	4	9	9.8	10	Numbers per square metre 9	12	3.7	2.3	Per cent. frequencies 65	71	33	30
Minesing swamp												
18			25	25			7	5			6	2
Areas in square metres			5	5	Numbers per square metre		1.4	1	Per cent. frequencies		24	8
Balsam lake												
5	16		25	25	22		9	9	9		6	7
Areas in square metres 3.2		5	5	Numbers per square metre 7		1.8	1.8	Per cent. frequencies 56		24	28
Dorset												
19			25	25			1	0			1	0
20			25	25			1	1			1	1
Totals..			50	50			2	1			2	1
Areas in square metres			10	10	Numbers per square metre	0	2	0.1	Per cent. frequencies		4	2

showed a peak in 1933. On the basis of this and the censuses, it is thought that hares reached a peak of abundance in 1933 and had become scarce by 1934.

The scatology indicated populations of 200, 300, 60, 30, per square mile for the four years 1932 to 1935. The censuses of hares frequenting the area about the cabin (table 4) correspond to populations of 800, 900, 100, 100, for the same years. Comparing these results it is evident that either the guess as to the area from which these animals were drawn was about one-third as large as it should have been, or there was a special concentration of hares in that locality. The latter alternative is believed to be nearer the truth.

(c) Frank's bay on lake Nipissing is in the same general region as Algonquin Park and showed much similarity in the

TABLE 7.—Estimated populations of varying hares at Buckshot lake.

| | Numbers per square mile | | | | |
	July, 1932	Sept., 1932	1933	1934	1935
In swamps alone	3,300	950	70	80	30
In the whole country	3,400	250	30	0.3	0

hare situation. The per cent. catch in trapping declined in 1933 and was very low in 1934 and 1935. In 1933 the same area was trapped and the traps set in the same places as in 1932, so that the low catch in 1933 might have been due to insufficient restocking from surrounding territory. New traps were added on an entirely different area the next year but the catch was low, indicating a true scarcity in 1934 and 1935. The censuses were based on the trapping and observation on the area trapped in 1932 and subsequently, namely, plots 2, 3, and 4. They did not include the new trapping areas added in 1934. It follows, therefore, that the censuses showed the influence of insufficient restocking in 1933, if such insufficiency really existed. In spite of this, the censuses indicated an abundance of hares in 1932 and 1933, followed by scarcity in 1934 and 1935. The numbers of hares observed per fifty miles of travelling were independent of the trapping and showed a peak in 1933. Hare droppings were numerous in 1932 and 1933, but scarce in subsequent years. Of the 1933 counts the small part that came from an untrapped

area, plot 16, yielded a high figure.   From all this it is plain
that 1933 was the last year of real abundance at Frank's bay,
just as it was in Algonquin Park.

The densities of population estimated from the scatology
are in the first line of table 8.   To avoid the effect of the
partial depletion of the trapped area in 1933, the count of
droppings from the untrapped plot alone was used for that
year.

The estimate from the observations of hares, while walking,
tells much the same story as that from scatology.   The actual
censuses obtained correspond to populations of 600, 400 (the
depleted trapping area), 80, and 0, for the years 1932 to 1935.

(d) Unfortunately the writer was unable to work at
Smoky Falls on the Mattagami river in northern Ontario until
1935.   The trapping figure was high.   Most of the trapping
was done with box traps to catch the animals alive and it is

TABLE 8.—Estimated populations of varying hares at Frank's bay.

| | Numbers per square mile | | | |
	1932	1933	1934	1935
Estimated from scatology	500	700	70	100
Estimated from hares seen	700	1800	200	30
Averages	600	1200	100	70

possible that this method gives higher catches than would
snaring in summer.   The census was a high figure, about one
thousand hares per square mile.   Numerous hares were
observed; the average being six per fifty miles of walking.
Droppings were abundant on the forest floor.   Altogether the
field work demonstrated that the hares were present in large
numbers.   The only other occasion on which the writer has
seen hares in such abundance was at Buckshot lake in July,
1932.   Nevertheless the local residents unanimously assured
me that hares had been even more common the summer
before, that is, in 1934.

(e) There still remain to be summarized the results at
the stations where only a short time was spent yearly.   Unfortunately it was not expedient to visit all of these every
year, especially in the year 1933.   The Bruce peninsula was
the only place examined regularly.

Near Kettleby in York county the varying hare "was certainly more numerous during the winter of 1929-30 than previously" (Snyder, 1930). In the course of field work in Christmas week, 1931-2, five hares were caught and several broke away from snares in a cedar swamp in 122 trap-nights, that is, a 4 per cent. catch. This indicates a fair abundance at that time in this swamp area about the Holland marsh, which is practically isolated by cultivated land from the forested areas further north. It is not known whether or not the hares in this locality took part in the general cycle by suffering a decrease in numbers in the next year or two.

At Minesing in cedar swamp "recent" droppings were not numerous in 1934 and 1935.

In a cedar swamp inhabited by varying hares near the north-west arm of Balsam lake, the scatology indicated a moderate abundance in 1932, followed by a decrease by 1934.

Varying hares were abundant near Wren lake in Haliburton county in September, 1931, as several at a time were seen frequently during the evening. In July, 1932, one evening a large number of hares were seen along the road. In 1934 and 1935 they were scarce, judging by the droppings.

At the base of the Bruce peninsula in a cedar swamp back of the dunes of Sauble beach, the counts of droppings were high in 1932 and 1933, but were reduced in 1934 and 1935. A hare was seen here on August 19, 1933. Near the upper end of the peninsula there was a sharp peak of abundance in 1933, as shown by the scatology.

Discussion

The field work has filled in some details that clarify and explain the questionnaire results. For Buckshot lake the general story from the questionnaires, being an average for a larger area, is not quite so abrupt in the decrease, but shows much the same picture of decrease during 1932 and 1933 and scarcity since the end of 1933. In the winter of 1932-3 correspondents reported the abundance that obtained in the early part of the previous summer, which is the reason that the curve does not show a greater decrease that year; however, 62 per cent. of the correspondents noted the decrease that had taken place since early summer. The next winter correspondents had no abundance to report in the year 1933-4.

The questionnaires for northern Ontario north of the height of land district, have definitely marked the year 1934 for the peak of abundance and showed no sign of general decrease at the time they were returned in the winter and spring of 1935. A year later in the first quarter of 1936, all correspondents reported a definite decrease. The field work, carried out in the spring and early summer of 1935, at Smoky Falls, suggested that the major part of the decrease did not take place in the winter of 1934-5 or in the early summer of 1935, but must have occurred in the latter part of the summer, the autumn, or the early part of the winter of 1935-6.

The work at other field stations corroborated the information obtained from questionnaires.

The results of the field work have borne out the idea of a progression of the time of the peak of abundance across Ontario from the south-eastern part to Cochrane district in the north. At Buckshot lake the end of the time of abundance was in the summer of 1932; in Algonquin Park and at Frank's bay, 1933; no field station was in the next zone; and at Smoky Falls the abundance ended in 1935.

In the matter of actual hare populations, estimates have been made by four of our correspondents at times of abundance. In Manitoba, near Whitemouth, in the autumn of 1933, Mr. V. B. Latta said there were about five thousand present per quarter section (a quarter of a square mile) of bush land, *i.e.*, twenty thousand to the square mile. This may well be an over-estimate.

Sergeant H. U. Green estimated for the winter of 1932-3 in the Riding Mountain National Park, Manitoba, "in favoured localities average of 5 to the acre"; and for the year 1933-4, "7 to 8 to the acre in favoured localities". Seven per acre means about forty-five hundred per square mile.

Mr. J. Anderson told of a thousand hares being taken in 1932-3 in a "two mile square area", and next year reported them increased and abundant in the country north of lake Nipissing, Ontario.

In the vicinity of Kapuskasing, Ontario, Mr. J. Raeburn reported that "On Dec. 9, 1933, two men shot 32 hares in about 40 acres, and on Dec. 16th they shot 23 more in the same area". That is, a total of fifty-five hares were taken off forty acres, which means there were at least nine hundred hares per square mile.

It will now be of interest to look at the few estimates of the population of varying hares that have been made by previous observers, comparing them with the estimates that have been presented in this study. Seton (1909, 1928) said the hares were scarce when there was only one to the square mile, and abundant when there were one thousand per square mile. He made an estimate in 1886: "Near the Spruce Hill, at the edge of the poplar woods, Carberry, Man., I stood; and looking around, counted the rabbits within a radius of thirty yards. They numberd 11, and there must have been many that I did not see; so that 20 would be a safe number at which to put them; that is 20 to the acre. But, dividing it by 2 to allow for the sparser places, it would total 5,000 to the square mile." Soper's (1921) statement about northern and western Alberta in 1912, to the effect that on a tract of about thirty acres he found hundreds of rabbits, is probably only a way of saying they were at a high level of abundance.

The estimates obtained in the present study are of the same order of magnitude as the above statements, namely from one to between one and five thousand per square mile for the range from scarcity to comparative abundance. Mr. Raeburn's is particularly interesting, because it was in the year before the peak of abundance, and in the year following the peak, but before the main decrease, the writer was in the same district and found about the same number of hares per square mile.

MISCELLANEOUS LIFE HISTORY NOTES

In the present section is included information on various phases of the life history of the hare, accumulated during the course of the investigation. Some of these data have a connection with the main problem but in others the application is more remote.

Reproduction

Available records of litters of young hares, born and unborn, are listed in table 9a, while the other records of young ones, lacking the data as to numbers in the litters, are in table 9b. If there is no reference, they are published here

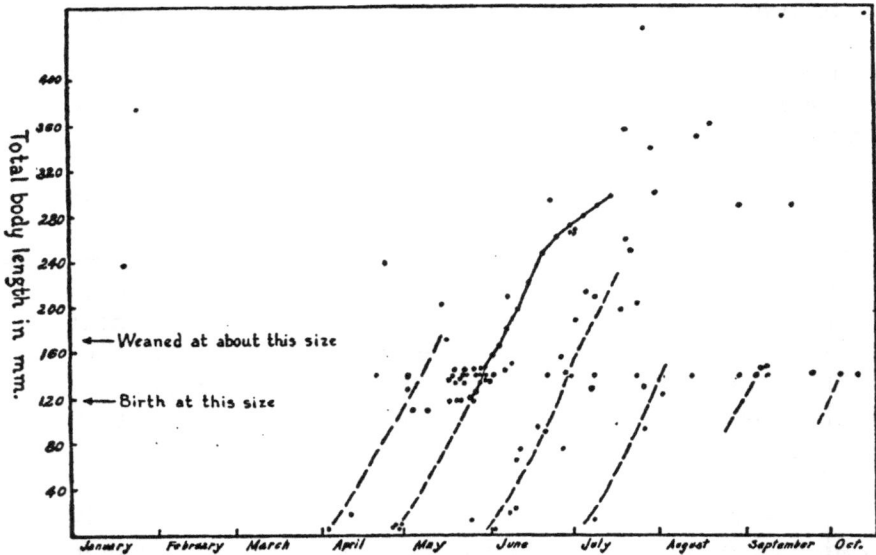

FIGURE 15.—Breeding of the varying hare in Canada east of the prairies. Each dot represents the size of a foetal or juvenile hare or the average size of the members of a litter; in some cases it refers to the collecting of a lactating female. The solid line joining a series of dots gives the averaged growth of three hares born in captivity. Broken lines point out the successive litters. The April litter does not occur in central or northern Ontario, or other parts of northern Canada. Dates to be read a week later than on scale, for northern districts, and a week earlier south of Muskoka. (Northern districts—above latitude of lakes Superior and Timiskaming.)

for the first time. This information has been plotted in figure 15, by size over date, ignoring the years. Each dot represents the length of a foetal or juvenile hare or the average size of the members of a litter. Only one dot was marked for a litter, no matter how many young were in it. Dots that refer to the collecting of a lactating female have been plotted at about the size of five days old young. Records from north of the level of Nipissing district have been plotted a week earlier than they occurred and those from south of Muskoka district a week later.   In the absence of any phenological contour map of the country, this was an approximate way of bringing the first litters into step.   The solid line joining a series of dots gives the averaged growth of three hares born in captivity.   These three, and some of the dots representing very small foetuses at the end of April, were known to be among the first litters of the season at Smoky Falls on the Mattagami river.   The foetuses and the growth curve have therefore been joined by a broken line, the whole

TABLE 9.—Reproduction of the varying hare.

Part (a).—Records of litters of young hares, both born and unborn.

Date	Locality	Observer	Data
	Ontario		
June 7, 1932	Frank's bay	author	2 young have been born
June 8, 1932	"	"	4 foetuses, sizes 22, 23, 25, 21 mm., average 23 mm.
June 9, 1932	"	"	3 foetuses, 64, 75, 56 mm., average 65 mm.
June 20, 1932	"	"	3 foetuses, 95, 90, 87 mm., average 91 mm.'
June 21, 1932	"	"	recently gave birth to 5 yg.
May 18, 1933	"	"	3 yg. recently born
May 23, 1933	"	"	3 foetuses, all 13 mm.
June 6, 1933	"	"	4 foetuses, 17, 18, 19, 20 mm., average 18½ mm.
April 27, 1934	"	"	2 foetuses, both 9 mm.
June 26, 1934	"	"	2 foetuses, 81, 71 mm., average 76 mm.
July 7, 1934	"	"	4 foetuses, about 15 mm. *and* suckling young
Aug. 29, 1935	"	"	3 yg. being suckled
May 17 to June 15, 1933	Smoky Falls on the Mattagami river	R. V. Whelan	1 embryo in each of 3 hares 2 embryos in each of 2 hares 3 embryos in each of 12 hares 5 embryos in 1 hare average 2.7 foetuses
May 3, 1935	"	author	2 foetuses, both 8 mm., mammary gland not yet developed

TABLE 9.—*Continued*

Date	Locality	Observer	Data
May 5, 1935	Smoky Falls	author	2 foetuses, both 7 mm., mammary gland not yet developed
May 6, 1935	"	"	2 foetuses, both 10 mm., mammary gland not yet developed
May 23, 1935	"	"	1 born (she ate it)
May 26, 1935	"	"	2 born (she killed them and ate all but head of one and chewed rump of other)
May 27, 1935	"	"	2 born (one lived)
June 1, 1935	"	"	2 born (both lived)
June 7, 1935	"	"	3 foetuses, each 6 mm. long. (she had borne young previously this summer)
July 14, 1935	"	"	2 young, lactating
May 13, 1931	Coboconk	J. Edmonds	3 foetuses
June 11, 1931	Jocko	A. F. Coventry	4 foetuses, about 75 mm. long
Summer, 1933	Port Arthur	A. Graesky	"about 3 and 4 young seen together"
July 1, 1932	Glen Orchard	E. Fairhall	"found 4 young hares in garden, well grown"
May 28, 1933	Timmins	C. Seguin	"litter of 4, about 2 or 3 days old"
1932	Mine centre	G. Stagee	"bunch of 7 young ones, only a few days old"
Summer, 1931	Gogama	J. G. K. McEwen	3 or 4 very young rabbits
July 1, Year?	Dorset	(Wright and Simpson, 1920)	young half-grown hares
July 24, 1924	MacGregor bay, Manitoulin county	B. A. Bensley	yg. ♂ T.L. 129 mm., and a twin same size
	Outside Ontario		
Aug. 1, 1932	Newfoundland	W. E. McCraw	"4 yg. in nest"
Jan. 19, 1934	Camrose, Alta.	F. L. Farley	"saw 2 yg. hares less than half grown"
April 15, 1933	Dauphin, Man.	H. U. Green	4 foetuses
April 17, 1933	"	"	4 foetuses
May 9, 1933	Lake Dauphin, Man.	"	3 foetuses
May 9, 1933	"	"	3 foetuses
May 9, 1933	"	"	4 foetuses
May 9, 1933	"	"	5 foetuses
May 9, 1933	Ochre river, Man.	"	5 foetuses
May 9, 1933	"	"	4 foetuses

TABLE 9.—*Continued*

Date	Locality	Observer	Data
May 9, 1933	Ochre river, Man.	H. U. Green	4 foetuses
May 9, 1933	"	"	3 foetuses
May 10, 1933	Dauphin, Man.	"	3 foetuses
May 10, 1933	"	"	5 foetuses
May 10, 1933	"	"	4 foetuses
May 10, 1933	"	"	2 foetuses
May 10, 1933	"	"	4 foetuses
May 10, 1933	"	"	4 foetuses
May 10, 1933	"	"	4 foetuses
May 10, 1933	"	"	6 foetuses
May 11, 1933	"	"	3 foetuses
May 11, 1933	"	"	5 foetuses
May 12, 1933	Lake Dauphin, Man.	"	5 foetuses
May 12, 1933	"	"	4 foetuses
May 12, 1933	"	"	5 foetuses
May 13, 1933	Dauphin, Man.	"	6 foetuses
May 13, 1933	"	"	5 foetuses
May 20, 1933	"	"	2 foetuses
June 5, 1925	Merrill, New York	(Seton, 1928)	5 young about 5 inches long, still nursing
June, 1905	Duluth, Minnesota	(Seton, 1909, 1928)	a doe followed by 4 young, half to two-thirds grown, and two yg. were about ready for birth
April 10, 1882	Carberry, Man.	(Seton, 1909, 1928)	2 small embryos
May 15, 1911	Alaska	(Dice, 1921)	3 foetuses about 110 mm.
June 24, 1911	"	(Dice, 1921)	6 large foetuses
August 2, 1911	"	(Dice, 1921)	6 large foetuses

Part (*b*).—Records of young hares and adult females lacking data as to numbers in litters.

Date	Locality	Observer	Data
	Ontario		
June (late), 1930	Frank's bay Lake Nipissing	J. M. B. Corkill	A small young rabbit
June 18, 1932	"	author	multiparous, but not lactating
July 23, 1933	"	"	yg. ♂ T.L. 202 mm.
July 23, 1934	"	"	lactating female
July 6, 1935	Smoky Falls on the Mattagami river	author	lactating female
July 11, 1935	"	"	multiparous but not lactating

TABLE 9.—*Continued*

Date	Locality	Observer	Data
July 12, 1935	"	"	yg. ♂ T.L. 214 mm., Wt. 215 gms.
July 28, 1935	"	"	yg. ♀ about 4 weeks old
June 2, 1933	Kapuskasing	F. N. Wiley	yg. 4 inches long.
Sept., 1930	Brent, Algonquin park	author	yg. about 9 inches long
Aug. 12, 1932	Biggar lake, Algonquin park	"	multiparous
Aug. 19, 1932	"	"	♀ T.L. 361 mm., Wt. 1.9 lbs.
July 30, 1932	"	"	♀ not lactating
July 30, 1932	"	"	♀ not lactating
July 19, 1932	Buckshot lake, Frontenac county	"	yg. ♂, T.L. 357 mm., Wt. 2.2 lbs.
July 26, 1932	"	"	yg. ♂, T.L. 443 mm., Wt. 2.3 lbs.
July 28, 1932	"	"	yg. ♂, T.L. 339 mm., Wt. 1.2 lbs.
Sept. 4, 1933	"	"	lactating
July 19, 1933	Arden	R. V. Lindsay,	yg. T.L., 260 mm., sex?
May 25, 1932	Denbigh	P. Plotz	yg. about 3 or 4 days old
Nov. 26, 1934	Metagama	M. V. Bates	nulliparous ♀, T.L. 429 mm., Wt. 3.4 lbs.
Jan. 24, 1935	Savanne	F. Edwards	yg. ♂, T.L. 375 mm., Wt. 1.3 lbs.
June 23, 1931	Timagami	A. F. Coventry	yg. ♀, T.L. 293 mm.
July 6, 1924	MacDiarmid, lake Nipigon	L. L. Snyder	yg. sex?, T.L. 265 mm.
Aug. 21, 1922	"	J. R. Dymond	yg. ♂, T.L. 350 mm.
July 30, 1925	Lowbush, lake Abitibi	....	yg. ♀ T.L. 300 mm.
July 7, 1925	Camp 33, lake Abitibi	L. L. Snyder	yg. ♂ T.L. 265 mm.
June 13, 1925	Lowbush, lake Abitibi	W. J. LeRay	yg. ♂ T.L. 210 mm.
July 24, 1925	"	"	yg. ♀ T.L. 200 mm.
July 15, 1933	Mouth of Kenogami river	A. G. Crisp P. W. Ferris	yg. ♂ T.L. 210 mm.
July 4, 1925	Lightning river, lake Abitibi	L. L. Snyder	yg. sex? T.L. 142 mm.
June 26, 1925	Ghost river, lake Abitibi	"	yg. ♂ T.L. 225 mm.
April 26, 1932	Gooderham	H. E. McColl	first yg. rabbits about 4 inches long, a week old

TABLE 9.—*Continued*

Date	Locality	Observer	Data
May 15, 1932	Boulter	H. Taylor	yg. about 1 week old
July 2, 1932	"	"	yg. several days old
May 21, 1934	Wako	J. A. Johnson	first yg.
May 28, 1933	South Magnetawan	J. M. McArthur	first yg.
April 20, 1933	Larson	A. E. Swaim	first yg.
June 4, 1932	North Branch	T. B. Brannan	first yg.
June 28, 1932	Fenelon Falls	K. Menzies	a young one
June 29, 1932	MacGregor bay	B. A. Bensley	3 generations of rabbits apparent around his cottage
	Outside Ontario		
May 15, 1933	Newfoundland	W. Nichols	first yg. seen ten days old
Nov. (middle), 1934	Quebec, P.Q.	R. Meredith	"saw some young and quite small hares very late"
Jan. 10, 1935	Nova Scotia	H. S. Cruikshank	yg. were ready to be born
Oct. (early), year?	Nova Scotia	R. W. Tufts	yg. "no bigger than my fist"
Oct. 4, 1933	Treesbank, Man.	S. Criddle	yg. lactating
April 2, 1934	"	"	"mating started"
Sept. 5, 1933	Whitemouth, Man.	V. B. Latta	yg. several days old
Sept, 1934	East of Red river and south of C. N. R. main line, Man.	P. A. O'Connor	"a few cases of very young hares observed in Sept."
April 24, 1933	Rennie, Man.	A. Hole	yg. hare at least 7 weeks old, snow on the ground
Oct. 10, 1932	"	"	a large hare suckling young
May 1, 1933	Riding Mountain National Park Man.	H. U. Green	first female nursing
April 17, 1933	"	"	female, no embryos
May 9, 1933	Ochre river, Man.	"	female, no embryos
May 9, 1933	"	"	female, no embryos
May 10, 1933	Dauphin, Man.	"	nursing
May 11, 1933	"	"	nursing
May 12, 1933	Lake Dauphin, Man.	"	nursing
May 12, 1933	"	"	nursing
May 13, 1933	Dauphin, Man.	"	nursing
May 18, 1933	Lake Dauphin, Man.	"	nursing
May 19, 1933	Dauphin, Man.	"	nursing
May 19, 1933	"	"	nursing
May 20, 1933	"	"	nursing

serving to indicate by rate of growth what dots on the graph
represent one seasonal group of litters.   Broken lines point
out the successive litters throughout the season.   The
majority of the hares start breeding at about the same time
in any one locality and so bear young at about the same date.
The April litter does not occur in central or northern Ontario
or other parts of northern Canada.   During the May reproduction time all females collected were pregnant.   There is
definitely a second major group of litters in June.   Frequently
a female has been taken that was both suckling young and
pregnant at this time.   The indications are that only a part
of the females have young in July.   There is no proof as to
whether or not individuals will raise three or more litters in
a season, but it seems very probable that some do have three
litters.   A few hares are born in August and September, and
it is interesting that the records fall into groups about a
month apart as do the early broods.

The average number of young per litter, from the data
of table 9a, was 3.4; the smallest number was one and the
highest number seven.   The relations of the rate of reproduction to the cycle of abundance are discussed in the section
on correlations of the cycle with other phenomena.

Growth

Three young have been reared in captivity.   Snowshoe
rabbit No. 154, which was captured on May 20, 1935, at
Smoky Falls on the Mattagami river, gave birth on the night
of May 26-7, to two young, one of which died accidentally
next day from catching a leg in the wire mesh of the cage.
At the same locality, rabbit No. 156 had two young on the
night of May 31-June 1.   Their total length (T.L.), tail
vertebrae (T.V.), hind foot (H.F.), ear length from skull, ear
length from notch and weight, were measured at intervals,
and are recorded in table 10.   The weight was measured in
grams until the beasts exceeded the capacity of the balance
and thereafter in pounds by a spring scale which was checked
and found to be as accurate as it was read.   Their growth in
total length has been averaged and plotted on figure 15,
mentioned earlier.

TABLE 10.—Growth of varying hares from birth (measurements in millimetres).

Date	Age in days	T.L.	T.V.	H.F.	Ear from skull	Ear from notch	Weight in grams
(a) Hare No. 157, born night of May 26-7, at Smoky Falls on Mattagami river							
May 28, 1935	2	122	12	37	22	20	49
May 31	5	142	13	42	25	23	75
June 2	7	157	14	44	29	27	97
June 3	8	163	16	48	31	29	106
June 5	10	168	19	51	33	30	127
June 6	11	167	18	51	35	33	132
June 7	12	178	21	56	36	34	133
June 8	13	177	19	53	38	33	141
June 11	16	191	19	58	37	35	159
June 16	21	217	24	65	46	39	199
June 21	26	238	25	73	50	45	250
June 26	31	266	25	79	56	49	333
July 4	39	283	24	91	59	54	454
July 12	47	301	24	97	61	59	1.1 pounds
July 23	58	326	26	100	66	59	1.2 pounds
(b) Hare No. 159, born night of May 31-June 1, at Smoky Falls							
June 1, 1935	0	118	9	34	20	18	48
June 2	2	122	11	36	21	19	48
June 3	3	126	13½	37	23	21	53
June 4	4	129	14	41	23	22	67
June 5	5	136	14	43	23	23	62
June 6	6	134	17	45	26	26	72
June 7	7	132	..	45	27	26	73
June 8	8	140	17	45	27	26	65
June 11	11	...	17	47	27	26	86
June 16	16	184	22	57	35	33	144
June 22	22	220	..	..	..	..	220
June 26	26	246	23	77	47	55	274
July 4	34	257	24	89	55	49	339
July 12	42	280	25	94	54	49	0.9 pound
July 17	47	290	..	..	..	..	0.9 pound
July 23	53	294	27	96	59	53	0.9 pound
(c) Hare No. 160, born night of May 31-June 1, at Smoky Falls							
June 1, 1935	0	119	11	36	21	18	58
June 2	2	130	12	37	22	19	58
June 3	3	132	14	40	25	22	70
June 5	5	141	17	43	28	27	77
June 6	6	151	17	46	27	26	85
June 7	7	151	17	48	27	26	91
June 8	8	159	18	49	30	28	103

TABLE 10.—*Continued*

Date	Age in days	T.L.	T.V.	H.F.	Ear from skull	Ear from notch	Weight in grams
June 11	11	...	18	53	34	31	115
June 16	16	193	22	62	39	36	166
June 22	22	222	..	..	..	..	231
June 26	26	243	25	77	53	46	277
July 4	34	259	24	84	54	48	365
July 12	42	271	25	89	58	52	0.9 pound
July 17	47	275	..	..	..	..	0.9 pound
July 23	53	278	27	94	59	53	0.9 pound

The sex ratio among the adult part of the population is fifty-three females per hundred of the population, based on all the adult hares for which records were available, *e.g.*, two hundred and sixty-nine.

Body Temperature

The body temperature of hares, while not a matter of life history but rather of physiology, may well be mentioned since some measurements were obtained. During the winter of 1932-3 two hares, No. 135 from Arden in Frontenac county and No. 136 from Timagami, Ontario, were in cages in the Hygiene Building, University of Toronto.

The rectal temperatures all were taken in the afternoon and are listed in table 11. At first the hares were excited at the strange procedure and their temperatures were abnormally high, reaching 104.4° F. However, they soon became accustomed to the handling, and their temperatures then fluctuated within about one degree total range. . The average of eight measurements on one hare and seven on the other, spread over six weeks, was 102.3° F. (the abnormal early measurements omitted).

Food

The food of the varying hare has been summarized by Richardson (1829), Seton (1909), Dice (1921), Lake States Forest Experiment Station (1935, 1936), and many pieces of information are scattered through the literature.

A plant that the writer has observed both wild and caged hares to refuse to eat is *Clintonia borealis*. A number of correspondents, including several Indians, have claimed that snowshoe rabbits show a preference for willow bark before dying off in numbers.

Hardy (1910) mentioned a hare eating fish, and Soper (1921) and Seton (1928) described their habit of eating flesh of carcases, even of one of themselves. One of our correspondents, Mr. James Anderson of Crystal Falls, Ontario, told that hares will eat the meat of deer carcases. Mr. G. H. D. Bedell of Dauphin, Manitoba, recorded dead rabbits

TABLE 11.—Body temperature of varying hares.

Date	Hare No. 135	Hare No. 136
Dec. 19, 1932	104.4	103.2
Dec. 20	103.6	102.8
Dec. 21	102.1	102.2
Dec. 22	101.8	101.6
Dec. 27	102.1	102.4
Jan. 4, 1933	102.9	102.3
Jan. 6	102.5	102.3
Jan. 8	102.7	102.4
Jan. 17	102.1	102.7
Jan. 30	...	102.1
Averages, omitting first two measurements	102.3	102.2

being eaten by the survivors. They will eat all kinds of frozen flesh according to Mr. W. H. Bryenton of Herb lake, Manitoba.

A food habit that has not yet appeared in the literature was observed by Miss Irene Hollingworth at Dorset in 1934, when she saw hares eat sand. Mr. R. V. Whelan of Smoky Falls on the Mattagami river wrote: "While trapping insects at night have observed hare within a few feet of my light eat sand for hours night after night." The author in July, 1933, watched a young hare of the season digging under the leached layer and eating the sandy soil at Biggar lake, Algonquin Park. Furthermore, sand has been found in the stomachs of several of the hares examined.

On a map of plot 21 at Smoky Falls on the Mattagami river all trees off which hares chewed the bark during the winter 1934-5 were marked. The girdled poplars were on the average about two and a half inches in diameter and the bark had been removed for an average of about two feet above the snow. A hundred and sixty-eight such poplars (*Populus tremuloides* and *grandidentata*) had been used, and in addition eleven small willows, fifteen alder (*Alnus incana*) stems, two white birch (*Betula papyrifera*) saplings, juneberry (*Amelanchier* sp.), and one jack pine (*Pinus banksiana*). It has been estimated in an earlier section that twelve hares were the population of this area, and ate the above quantity of bark in one winter. The average ration per hare per winter was fourteen poplars and a taste of the other foods.

Habitat

Richardson (1829) said of the varying hare: "In the northern districts it resides mostly in willow thickets. . . ." In Ontario Miller (1897) noted that it was common in swamps at Milton. Tamarack and cedar swamps, dry copses varied with open glades, and willow thickets were described as its preferred habitats in Manitoba, by Seton (1909, 1928). The above are representative early statements and many notes have appeared since.

To gain local information on habitat preferences, snaring was done on three adjacent plots in different types of vegetation at Frank's bay on lake Nipissing. Plot 2 was in spruce-larch swamp with wet sphagnum ground cover and many shrubs of several kinds, such as alder (*Alnus incana*) and mountain holly (*Nemopanthus mucronata*). Plot 3 was on the jack pine-lichen-covered rock ridge alongside the previous plot; and plot 4 was in mixed second-growth poplar-birch-white pine-red pine-balsam with much hazel (*Corylus rostrata*) shrubbery. The record of snaring in table 3 shows that no hares were caught on plots 3 and 4, except one that broke the snare and escaped on plot 4; all those collected were in the swamp. This result exaggerated the true situation because the greater use of visible runways in the swamp made it easier to set snares effectively there. Of the three hares observed during night observation, one was in the swamp and the others on its margin, barely inside the mixed bush.

In the winter of 1931-2 a question regarding the habitat of the hare was included on the questionnaire and the replies are summarized in table 12. Indefinite reports have been omitted.

The replies have shown that coniferous swamps, willow-alder swamps and thickets, and poplar-birch second growth on brulés and cut-over areas comprise the most used habitats

TABLE 12.—Habitat of the hare as shown by replies to questionnaires.

	Maritimes and Quebec		Ontario east of lake Nipigon		Ontario from lake Nipigon westward		Western Canada	
	Replies	Per cent.	Replies	Per cent.	Replies	Per cent.	Replies	Per cent.
Coniferous swamps	7	30	55	66	4	35	3	7
Willow-alder	6	25	4	5	0	0	17	40
Second growth cut-over and brulé and poplar-birch	7	29	16	20	5	40	11	25
Spruce forest, not swamp	1	4	0	0	0	0	6	14
Jack pine	0	0	1	1	3	25	5	12
Mixed forest	2	8	5	6	0	0	0	0
Hardwood forest	1	4	2	2	0	0	1	2
Totals	24	100	83	100	12	100	43	100

of these hares. The type of vegetation which is used most generally is influenced obviously by the relative frequency of occurrence of these types in a particular region. In the maritime provinces and Quebec, the data did not show one type as used more than any other of the three; but in Ontario as far west and north as Algoma, the coniferous swamps of cedar and spruce chiefly were the "typical" habitat beyond all question. Westward from lake Nipigon in Ontario, jack pine covers a larger part of the forest area than in the rest of the province (Sharpe and Brodie, 1931) and was mentioned as an important habitat for hares. In western Canada the coniferous swamps rarely occur, their place being taken by willow-alder swamps and thickets, which were reported as the chief habitat of the hares. Spruce forest, not swamp, and jack pine forest are both important in this region. Young poplar-birch bush is much frequented by the hares in all parts of Canada.

For the region about Buckshot lake in the north part of Frontenac county estimations of density of population in the swamps and in the whole of that region were given in table 7 and are quoted here in the first and fourth lines of table 13. The amount of swamp in the region studied about Buckshot lake is 5.4 per cent. of the land area. (The report on which this figure was based was lent through the courtesy of the Ontario Forestry Branch.) The 5.4 per cent. figure was

Table 13.—Density of population in different habitats.

	July, 1932	Sept., 1932	1933	1934	1935
Numbers of hares per square mile of swamp	3,300	950	70	80	30
5.4 per cent. of above; = numbers in the swamps of average square mile	180	50	4	4	1½
Numbers per square mile of country outside swamps	3,200	200	25	0	0
Total for the country	3,400	250	30	4	1

obtained by adding the figures for the following three types: black spruce-balsam-black ash-cedar, alder swamp, and coniferous swamp. This applied to the southern and Mazinaw tracts of the Eastern Provincial Forest, which includes much of the area concerned in this field work. Therefore, 5.4 per cent. of the densities of population of hares in the swamp types should be the actual numbers of hares in the swamps of the average square mile of the country (second line of table 13). By subtracting those in the swamps from the totals for the whole country, the numbers in the rest of the country outside of the swamps were obtained (third line of table 13).

It is shown that hares reached as great or greater density of population in the rest of the country, when they were near their peak of abundance in July, 1932, as they did in the swamps. By September of that year they had decreased to one-sixteenth of their former numbers outside swamps, but only to one-third in the swamps. During the next year the decrease was to about one-tenth in both habitats; but for the succeeding two years, 1934 and 1935, the only hares in the country were in the swamps. In these swamps the hares apparently found the most suitable conditions for survival. It appears, then, that the cedar, spruce, and alder-willow swamps are the typical or optimal available habitat of the hares in the Buckshot lake region of Frontenac county.

Territory and Runways

No sign of use and defence of a definite area or territory per animal or pair was seen among hares; their movements overlap indiscriminately. On the map of plot 21 the rabbit runways, that is, paths travelled by more than one set of tracks, were marked, and they interconnect completely; in fact a main "highway", heavily tramped down, may be traced from near the south end of the plot three-quarters of the way across it. The runs chiefly join food tree to food tree or to a hole or burrow. The greater concentration of runways near the meadow is to a small extent only apparent, due to the fact that the snow was beginning to thaw daily by the time the survey reached the northern part of the plot; but the concentration was mainly due to the presence of alders around the margin of the meadow, and denser cover of small conifers and poplars in the neighbouring bush than in the northern part of the area. After a fresh snowfall most of the same runways would soon be evident again. At Frank's bay and other places, the same runways through the swamps were found each summer. They have a permanence because the food and burrows and other features influencing the daily life of the hares remain in about the same locations.

Range of Movement

Seton (1909, 1928) estimated the home range at twenty or thirty acres in dense woods and perhaps twice as much in open woods; and gave instances to show that some individuals

will pass their lives within a radius of two hundred yards.
A few observations made in the present work may be mentioned; nine hares were released with aluminum tags on their
ears at various times and places and two returns were secured;
one at Biggar lake in Algonquin Park was seen the next year
in the same place, and the other, at Smoky Falls, was seen
three months after release two hundred and fifty yards away.
At Frank's bay in 1934 a hare broke a snare and was caught
in another snare a hundred and ten yards away with the first
noose still around it.   At Smoky Falls three times the track
of a hare was traced for a considerable distance on the snow;
one was lost after it had travelled a hundred and seventy
yards and was a hundred and fifty yards from its starting
point; a second was lost after it had travelled two hundred
and sixty yards and was sixty yards from the hole out of
which it had come for the evening's activity; the third was
lost after it had travelled three hundred and forty yards and
was eighty yards from the hole it had left.

"Forms" and Burrows

It has long been the accepted belief that "The American
Hare does not burrow" (Richardson, 1829), and that, an
assumed corollary, it does not live in, or even enter, holes or
dens, even ready-made ones.   However, Seton (1909, 1928)
did notice that a hare on his grounds had the habit of staying
in a certain rocky crevice for several days after each heavy
snowstorm.   What Seton said is still true, namely, that no
one (to our knowledge) has actually examined and reported
on the nest in the wild, containing young of the snowshoe
hare.   Bachman (1851-6) had two captive hares which produced young in captivity in "a nest of straw, the inside of
which was lined with a considerable quantity of hair plucked
from their bodies".   The two that brought up young ones at
Smoky Falls during the present study omitted the hair,
accepting simply the grass and piece of burlap that was
supplied.

Of Alaska, Dice (1921) wrote: "In winter they sometimes
make their forms under branches laden with snow, but they
apparently never burrow into the snow.   They usually use

an uncovered form which is only partly protected on the sides."

An occurrence that ran counter to the above was described by Snyder (1930) in these words: "One specimen was snared by the writer at the entrance of a burrow beneath a brush-pile in the depths of the swamp. It could not be determined what kind of animal had made the burrow or whether it was a natural or accidental hole leading into the ground but it was established that the hare will resort to underground protection during the daytime when the opportunity is afforded."

It has been evident to the writer since the first season's field work that the whole story has not yet been told, at least for Ontario.

At Smoky Falls a hare disappeared from sight behind a shrub and could not have run away without being observed; the entrance to a hole through actual soil, 3×4 inches in size and deeper than arm's length, was concealed under the low branches of a balsam seedling beside the alder shrub and a stump. A box trap was set there and next morning it contained a hare of the same whitish-grey appearance (date May 20, 1935) as the one seen the day before.

In winter tracks showed that hares entered the numerous holes in the snow beside stumps, fallen trees (fig. 13), bent over alder branches, low balsam or spruce branches, and even, rarely, a hole in bare smooth snow. Thirty-eight holes were used by hares on plot 21 at Smoky Falls. The census showed twelve hares on the area, which allotted three holes per hare. From tracks in fresh snow it was seen on more than one occasion that a hare came out of one hole and presently entered another. Many of these were holes only under the snow, but twelve of them, one per hare, continued underground, and were used through the summer. The hares certainly made and kept open the holes through the snow, but they have not been seen digging holes in the ground. Whether they do, or simply appropriate what holes already exist, remains an open question. One of our correspondents, Mr. T. Raycraft, said: "They do not use 'forms' as nests but live in burrows. They utilize ground hog holes and some hares dig holes themselves." The best places to set traps

,were beside the holes.   Possibly a shortage of suitable holes would limit the population in certain habitats.

No "form" was observed on this plot but a mile away a hare was seen in daytime lying in a saucer-shaped depression in the snow on top of a low overturned stump.   At Frank's bay on lake Nipissing two "forms" were found (one by Dr. C. H. D. Clarke) in summer from which hares were flushed; both were patches of sand on the sunny side of a tree.

Nocturnal Habits

Only a very few of the hares in the woods were seen or flushed in daytime; they were in holes.   Varying hares are largely nocturnal in their habits (Seton, 1909, 1928), and have shown, in the present field study, a decided decrease in activity after about eleven or twelve o'clock followed by a small increase in the early morning.

BACTERIOLOGY OF THE VARYING HARE

Specimens of hares were examined bacteriologically since it was thought possible that the decrease in numbers might be caused by an epizootic bacterial disease.   That fine naturalist, Richardson, wrote (1829): "At some periods a sort of epidemic has destroyed vast numbers of hares in particular districts. . . ."   Darwin (1859) suggested the same thing in more general terms; to quote: "When a species, owing to highly favorable circumstances, increases inordinately in numbers in a small tract, epidemics—at least this seems generally to occur, with our game animals—often ensue; . . ." Many observers have mentioned epidemics of disease that they assumed must have taken place in order to bring about the observed decimations of wild animal populations; numbers of dead animals were actually found in some cases.   Elton (1931) summarized the published records of epidemics, that is, heavy and sudden decreases in population, of some thirty-six kinds of mammals, of which only a few had been studied bacteriologically.

Epidemics occurring among animals in a state of nature have been traced to definite bacterial causative agents in only a small number of cases. It will be pertinent here to review these recorded epidemics.

The causative organism of bubonic plague, *Pasteurella pestis* (Yersin and Kitasato, 1894) Bergey *et al.*, was isolated independently and almost simultaneously by Kitasato (1894) and by Yersin (1894) from rats at Hong Kong. They recognized and proved the identity of plague in rats and in man. Clemow (1900) suggested that plague in Mongolia was essentially a disease of rodents, mentioning marmots (*Marmota*, formerly *Arctomys*) in particular. Quoting Topley and Wilson (1931): "Numerous workers, particularly Ogata (1897), Simond (1898), and Gauthier and Raybaud (1902, 1903) were responsible for showing that the disease is primarily one of rodents, and that it is spread to man by the agency of infected fleas. This conception was criticized by several workers (Nuttall 1898; Galli-Vallerio 1900, 1903), but was definitely proved to be correct by the English Plague Commission (Report 1906)." Ground squirrels (*Citellus*) were convicted as a reservoir of plague in California by McCoy (1910). Wu Lien-Teh (1924) after some ten years' work proved that marmots were one of the most important reservoirs of bubonic plague in Mongolia, as had been suggested by Clemow above. A summary of the evidence associating plague epidemics of humans with fluctuations in the populations of wild animals was published by Elton (1925); new information was added and the list of animals that had been found dying of bubonic plague brought up to date by Pirie (1927).

Danysz (1893, 1900) obtained cultures of "mouse typhoid" bacteria from epidemics among field mice, and this or other similar bacteria, usually *B. enteritidis*, have been used repeatedly in Europe to initiate epidemics among wild mice that were becoming overabundant (White, 1929). The success of these artificially generated epidemics is questionable, however, because the mice would usually be on the point of experiencing some natural epidemic in any case. Christy (1892) after seeing hundreds of rabbits lying dead in Manitoba wrote: "Why cannot we find in this disease a means of combating the Rabbit-pest in Australia?" The destruction of rabbits by means of the microbes of chicken-cholera was advocated by Pound (1897) in Australia.

An epidemic among caged field mice (*Microtus*) was recorded by Loeffler (1892), and the causative agent named *Bacterium typhi murium*. Topley initiated a study of experimental epidemics among caged mice in 1919 (Topley, 1919; Greenwood and Topley, 1925).

Horne (1912) obtained evidence that epidemics among lemmings in Norway during their periodic emigrations were of bacterial origin. He discovered bacteria, *B. pestis-lemmi*, which were fatal to guinea pigs and other animals. Another micro-organism, "*Streptothrix lemani*", was found by Johan-Olsen to be commonly associated with a skin disease from which the lemmings suffered (Collett, 1895). It is altogether likely that these findings do not represent the whole story and that further investigation would throw new light not only on the lemming situations but on the general problem of population control.

Martelli (1919) recorded enormous mouse (*Pitymys*) plagues in Italy, notably in 1916, when the mice died from epidemics of bacterial origin.

House mice (*Mus musculus*) and field mice (*Microtus*) were found to be dying of an epidemic, which Wayson (1927) found to be caused by an organism which he called *B. murisepticum*.

The South African gerbille (*Tatera lobengula*) suffers from epidemics of disease caused by the Tiger river bacillus, *Listerella hepatolytica*, and also from bubonic plague (Pirie, 1927); epidemics in 1928 due to a new species of *Pasteurella* were recorded (Mitchell and others, 1930).

Keane (1927) reported an outbreak of foot-and-mouth disease, which is caused by a filtrable virus, among deer in the Stanislaus National Forest in California.

Rocky Mountain Spotted Fever occurs among rodents and other wild animals and has been transmitted to man by wood ticks (Parker, 1929; Day and Shillinger, 1935).

Epizootic fox encephalitis (Green, 1925; Green, Ziegler, Green and Dewey, 1930) is a disease caused by a filtrable virus which has been responsible for heavy losses to fox ranchers, and Green (1931) expressed the opinion that the disease occurs among foxes (*Vulpes fulva*) in the wild state. Elton (1931) recorded epidemics among sledge dogs and arctic foxes (*Alopex lagopus*) of a disease resembling epizootic fox encephalitis.

Fox paratyphoid .has occurred in epizootic form on fur ranches and it seems probable that it occurs in the wild (Green, 1925; Shillinger, 1933).

A disease affecting moose (*Alces americana*) has been described by Thomas and Cahn (1932), Wallace, Thomas and Cahn (1932), Cahn, Thomas and Wallace (1932), and Wallace, Cahn and Thomas (1932); the causative agent being the bacterium, *Klebsiella paralytica*.

Bang's disease (contagious abortion) exists among elk and buffalo on big-game preserves (Shillinger, 1933).

The common rat (*Rattus norvegicus*) acts as a reservoir of infection for the disease, endemic typhus fever (Day and Shillinger, 1935).

Day and Shillinger (1935) described an outbreak of rabies among coyotes (*Canis nebrascensis*) in New Mexico, and another outbreak among foxes in Maine.

At Cos Cob, Connecticut, in 1903 and 1904, nearly all the rabbits and hares of several species with which the estate was stocked died with similar symptoms; and the medical report on the one that was examined was to the effect that it died of septicaemic infection, *Staphylococcus aureus*, the common germ of boils (Seton, 1909, 1928). In 1906 a snowshoe rabbit died there of "tuberculosis of the liver".

The disease, tularemia, has been left to the end of this review of the literature on epidemics among mammals in nature, because it brings us back to hares and rabbits. A "plague-like disease" was described by McCoy (1911) from forty-two naturally infected ground squirrels (*Otospermophilus grammurus*), in Tulare county, California. McCoy and Chapin (1912a, 1912b) discovered the causative organism, naming it *Bacterium tularense* after the county whence came the infected ground squirrels. They furnished serological proof of the infection of two laboratory workers. Several doctors noticed a disease which became known in Utah as "deer-fly fever", and was described by Pearse (1911) under the title "insect bites". The organism was found in a cottontail rabbit (*Sylvilagus floridanus*), and a case of infection in man was confirmed by the isolation of *B. tularense* (Wherry, 1914a and b; Vail, 1914; Wherry and Lamb, 1914).   Francis (1919) showed that "deer-fly fever" was an infection by *B. tularense* and he named the disease "tularemia".   In Bergey's "Manual of Determinative Bacteriology" (1930) *B. tularense* is listed as *Pasteurella tularensis*.

McCoy and Chapin (1912a, 1912b) passed the infection from animal to animal by the flea (*Ceratophyllus acutus*).   It was experimentally proven that the disease was transmitted by horse-flies (*Chrysops discalis*) (Francis and Mayne, 1921); by wood ticks (*Dermacentor andersoni*) (Parker, Spencer and Francis, 1924); by the rabbit tick (*Haemophysalis leporis-palustris*) (Parker and Spencer, 1925); from infected female ticks (*Dermacentor andersoni*), to their progeny (Parker and Spencer, 1926); by the wood tick (*Dermacentor occidentalis*) (Parker, Brooks and Marsh, 1929) and by the eastern wood tick (*Dermacentor variabilis*) (Green, 1931).   The disease can be acquired through handling infected carcases.   Lake and Francis (1922) showed that the organism can pass through the intact skin of guinea pigs.

Many animals, including all the rodents that have been tested, and some birds (including the ruffed grouse, *Bonasa umbellus*), are susceptible to tularemia.   In geographic distribution it has been found in the United States, Japan (Francis and Moore, 1926), Russia (Nikanarox, 1928; and Roubakine, 1930), Norway (Thiötta, 1930, 1931), and in Canada, a human case in Ontario (MacNabb, 1930); and (Hudson, 1930), and a snowshoe rabbit in British Columbia (Parker, Hearle, and Bruce, 1931).

Francis (1925) found seventeen jack rabbits infected with tularemia out of a total of five hundred and fifty-six shot in a stretch of country sixty miles in length in Utah in 1920.

A marked decrease in abundance of cottontail rabbits took place in the lake Alexander area of Minnesota in 1932 and Green and Shillinger (1932) found ten rabbits infected with tularemia out of twenty-three examined.   The incidence of *Pasteurella tularensis* in wood ticks in that area had been low and decreasing during the years 1929, 1930, and 1931. In their next report (1934) a rapid increase in the percentage of infected ticks was found in the fall of 1932 and during 1933. The disease was present in 1933 among snowshoe rabbits, but no appreciable mortality developed. Between the spring census of 1933 and that of 1934, there was a decrease in the population of snowshoe hares, while tularemia was found to have increased among hares and ticks (Green and Shillinger, 1935; Shillinger, 1935). They concluded: "It is only reasonable to ascribe to the observed spread of tularemia an important role in these losses."

Five of the above diseases are recognized as serious for humans and several others have been known to cause human infections. The study of diseases of wild animals has thus a direct importance to public health. Indirectly, it is valuable for the light that may be thrown on the mechanism of epidemics in general.

Methods

In taking intestinal cultures from living hares fresh droppings were sometimes cultured in plain beef broth, but at other times a thermometer, after being used for measuring rectal temperatures, previously sterilized in 10 per cent. phenol and washed in alcohol and sterile saline, was rinsed in sterile beef broth. The broth was incubated for one day at about 37° C. and spread by a platinum loop on culture plates of MacConkey's medium (MacConkey, 1900) and of plain beef agar. Strains were cultured from single colonies on the plates, selecting as great a variety of colonies as possible from both plates. These strains were put through the various tests required for identification after picking single colonies several times to get pure cultures. At autopsy intestinal cultures were taken by means of sterilized Wright's pipettes; which are of glass with the end drawn out to a capillary tube and sealed in a flame, a plug of cotton wool being in the other end. The surface of the intestine was first seared with a hot piece of metal. Sterile broth was drawn into the pipette by a rubber bulb, after breaking off the sealed end. The broth was injected into the organ through the seared surface, mixed about, withdrawn, and discharged into the tube of broth. Aseptic precautions, as flaming the pipette and tube, were observed throughout. Bacteria collected in the field

work were brought back to Toronto for identification. After
a sufficient number of cultures had been examined to give an
idea of the "normal" flora, only those intestinal cultures
which were colourless on MacConkey's agar were saved for
identification and study. Such strains do not ferment lactose
and are suspected of being pathogenic until proved otherwise.

A few cultures were made from the nose and throat by
sterile cotton swabs. From the organs, such as spleen, liver,
or lungs, cultures were taken when desired by Wright's
pipettes used as described above. A method used occasionally, however, was to rub a piece of the organ over the
surface of the culture medium. There is greater danger of
contamination with this process, but it can be sometimes used
satisfactorily. Blood cultures were obtained from the heart
either by a sterile syringe or Wright's pipette.

For serological tests blood was obtained from the heart
at autopsy if the specimen was fresh enough, or if the hare
were alive, from the lateral vein of an ear. When in Toronto
an electric centrifuge could be used to separate the cells from
the serum, but in the field the blood was allowed to clot,
"rimmed", i.e., freed from the sides of the tube with a sterile
wire, left to shrink and settle, and the serum drawn off with
a Wright's pipette. In the field kit carried in a rucksack
were forceps, scissors, scalpel, an old scalpel blade for searing
surface of organs, syringe wrapped in brown paper and
sterilized, sterile Wright's pipettes, sterile tubes with rubber
stoppers, some empty and some with dextrose beef broth in
them, and a small alcohol lamp. The bacterial cultures and
blood samples were taken immediately when hares were shot
in the woods.

In the field work a chicken incubator with a coal oil
burner and hot water pipes was used for culturing bacteria.
An alcohol lamp took the place of a bunsen burner and a small
single burner coal oil stove boiled the water for sterilizing
syringes and instruments and melting agar culture media.

Guinea pigs were taken to field stations for experimental
injections with tissue or pure cultures of bacteria and for
supplying blood for culture media. Usually blood could be
drawn asceptically from the heart of a living guinea pig by a
syringe and needle repeatedly at intervals of about a week.

Both domestic rabbits and varying hares were used for experiment in the field. The cages in which animals were kept have been described in the section on methods of field work.

The "Normal" Flora of the Varying Hare

The aerobic bacteria that occur commonly without causing apparent harm to the host rabbit were isolated and identified. The work was confined to the aerobic bacteria of the intestine, except for only a few cultures from the throat or other parts. Cultures were taken at intervals of about two weeks from two hares that were kept in captivity in the Hygiene Building during the winter of 1932-3. They were fed chiefly on natural foods, such as cedar and poplar twigs, with the addition of some oats and carrots and water. Other hares, fresh from the woods, have been used as opportunity occurred in order to get a more representative sample. The identifications were made according to the descriptions and keys in Bergey's "Manual" (1930). Ninety strains, isolated from a total of twenty-three hares, are summarized in table 14. Some, or possibly, many, of these "normal" inhabitants of the intestine might be capable of causing disease under conditions that predisposed the animal to their attack, but they have all been found without any sign of disease. Some of the identifications are provisional (marked with a query), and will be tested further. The Salmonellas must be checked by agglutination tests.

Certain of these strains which one might suspect of causing disease have been tested but pathological symptoms did not develop.

Salmonella suipestifer (Kruse) Lignieres

Isolated from a "normal" hare No. 64 from Arden and a second strain from hare No. 112 from Frank's bay on lake Nipissing. It caused no disease in guinea pigs on subcutaneous injection, nor in a varying hare on feeding.

Salmonella abortivo-equinus (Good) Bergey *et al.*

Isolated from spleen of a guinea pig which had died three days after injection of spleen material from an apparently normal hare No. 114 from Frank's bay on lake Nipissing. Injected subcutaneously into a domestic rabbit the culture occasioned a red swelling which cleared away in ten days. Fed to a varying hare with the food it caused no harm.

Pathogenic Bacteria

The bacteria that have been found to cause disease are as follows:

Pneumococcus Type XIX or *Diplococcus pneumoniae* Weichsel-baum in Bergey's "Manual" (1930)

TABLE 14.—"Normal" flora of the intestine of varying hare (aerobic bacteria only) cultures from twenty-seven hares.

Name	Number of strains	Per cent. frequency
Streptococcus faecalis	1	1
Staphylococcus albus	3	3
Escherichia coli	16	18
Escherichia communior	37	41
Escherichia alcalescens	6	7
Aerobacter aerogenes	11	13
Aerobacter cloacae	1	1
Proteus asiaticus	2	2
Proteus vulgaris	1	1
Salmonella suipestifer	2	2
Salmonella abortivo-equinus	1	1
Eberthella enterica	2	2
Alcaligenes sp.?	1	1
Bacillus several species	6	7
Total	90	100

Ten ticks off varying hare No. 153, at Smoky Falls on the Mattagami river, macerated and mixed with saline, were injected into a guinea pig, which steadily sickened, became thinner, the fur became scraggly and stood on end, and the animal died on the thirteenth day. The only visible effect was an enlargement and congestion of the spleen. Cultures on plain agar were sterile. The original hare remained well. The spleen of the guinea pig was injected into a domestic

rabbit subcutaneously and it died in four days, with hemorrhagic vessels and necrosis of the muscles of the leg under
the point of injection, a yellow colour in the connective
tissues, lungs congested, and spleen enlarged to twice its
normal size. The blood contained an abundance of dip-
lococci, and cultures on guinea pig blood agar were successful.
No colour effect was noted on this blood, but better growth
was obtained on chocolate blood agar, producing a zone
of green.

A guinea pig injected with lung material from the above
rabbit died on the fourth day, with a cavity under the point
of injection and around it a mass of blackish-grey firm tissue,
but no other marked lesions. Possibly the greater severity
of the infection in this pig, compared with the others used,
was due to the pneumococci having come from the lung,
the typical habitat of pneumococci.

An emulsion of living pure culture injected subcutaneously
into varying hare No. 154 caused its death in four and a half
days. The same bacteria were recovered in culture from it,
fulfilling Koch's Postulates (as summarized by Loeffler, 1886),
which constituted proof that the organisms caused the observed disease. As in the other rabbit, the connective tissue
of the whole thigh was yellow, the muscles swollen, necrotic,
firm, white, with hemorrhagic vessels, retroperitoneal glands
much enlarged and congested, spleen enlarged to three times
its normal size, firm and congested.

A guinea pig died on the twelfth day after injection with
a pure culture.

Cultures were sent to the Connaught Laboratories as it
was feared the strain might be lost if kept in the field for
several months. Miss E. A. Anderson identified them as
Pneumococcus Type XIX.

This fatal pathogenicity and muscular necrosis on subcutaneous injection was an addition to the knowledge of
Type XIX *Pneumococcus*. This strain differed in degree of
pathogenicity from the standard stock at the Connaught
Laboratories. A rabbit injected with that standard strain,
which had been passed through mice regularly, developed a
stiff leg but recovered completely.

As to the source, it apparently came from the ticks

(*Haemaphysalis leporis-palustris*), but it is not impossible that the first guinea pig acquired it otherwise. My own throat did not carry pneumococci in 1933 nor in the fall of 1935, some months after the culture was found, so that that source is not definitely ruled out.

Staphylococcus aureus Rosenbach (1884)

Seton (1909, 1928) lost some hares and rabbits through an epidemic and one animal that Dr. Seelye Little examined for him died of "a septicaemia due to infection by the staphylococcus *Pyogenes aureus*". The rabbit had ulcerated sores on the skin and enlarged lymph nodes, but no involvement of the heart, lungs, or abdominal organs.

In the present study this disease has been found frequently. The causative agent, the common pus-forming germs of boils and suppuration in wounds, is now known (Bergey's "Manual", 1930) as *Staphylococcus aureus* Rosenbach (1884). The disease was usually seen as only a few localized masses of cheesy pus surrounded by walls of necrotic tissue, or the later abscess stages, such as furuncles or carbuncles. Osteomyelitis, or suppuration and necrosis of bone, was not found. Cases occurred in hares at the following localities: Arden, Bradford, Coboconk, Fergus, Havelock, Kapuskasing, Key Junction (in Parry Sound), Markdale, Metagama, two from Oba, Smoky Falls on the Mattagami river; and in diseased cottontails (*Sylvilagus floridanus*) from Bowmanville and King (York county).

The usual source of infection appeared to be through wounds. Several hares had a furuncle on a hind foot, which would be a likely place for small cuts. The animals scratch their neck and head frequently, which probably accounts for the common occurrence of sores about these parts. The infection in the hare at Smoky Falls was initiated by a young leveret being weaned too late, so that its teeth wounded the teats which became infected. A domestic rabbit at that place was bitten repeatedly by another rabbit and developed infection.

The only fatal case was of a sluggish hare that Mr. J. Raeburn caught by hand on October 8, 1934; it died the next day, and he shipped it to us, hare No. 119. The lesions could be summed up as an empyemia. Almost all the important lymph nodes in the body were enlarged and had undergone cheesy degeneration. The lungs contained many pus pockets and the pericardial cavity was filled with nodules containing pus; the heart muscle itself was degenerated and riddled with pus pockets. Cultures from blood, spleen (which contained a mass of pus), pericardial mass, and from an enlarged parotid gland all gave *Staphylococcus aureus* only. Various media were used and both anaerobic and aerobic incubation, but with the same result, except that no bacteria grew anaerobically. Smears from the enlarged glands, *etc.*, were stained with the Ziehl-Nilsen carbol fuchsin acid fast method but no tubercle bacilli were found.

Two guinea pigs and a domestic rabbit were injected with material direct from the dead hare; the former did not become sick, but the latter weakened and developed masses of pus at the site of injection. It became paralysed in the hind quarters in a month and showed violent peristaltic motion and some incontinence, and died a month later without recovering from the paralysis. Dr. C. E. Dolman said the symptoms, excepting the swellings, were typical of a sublethal dose of staphylococcus toxin. Two rabbits were injected subcutaneously with pure cultures of the bacteria and in a month noticeably large soft masses of pus had formed which were still slowly growing in size at the end of five months. The original strain was recovered from these.

Dr. C. E. Dolman of the Connaught Laboratories, who had studied staphylococcus toxin and the treatment of staphylococcic infections with toxoid and antitoxic sera (Dolman, 1933, 1934), tested the above strain. The finding was that it was non-haemolytic, strongly yellow-coloured, and produced only a small amount of toxin, less than forty haemolytic units. He and his successor, Dr. J. S. Kitching, measured the antitoxin content of the sera of nine hares, including the fatal case, No. 119, all of which had less than one-tenth of one standard unit per c.cm., which is practically none.

The fact that varying hares have almost no antitoxin wherewith to neutralize staphylococcus toxin goes far toward explaining the frequent infections and the inability to over-

come them.   Even open skin abscesses have not healed in
the cases that have been watched, some of them for three
months.

Pasteurella tularensis (McCoy and Chapin, 1912) Bergey et al, 1912

No hares were found sick with tularemia, but proof of
its occurrence is detailed below.   Macroscopic agglutination
tests were made with the blood sera of hares and antigen of
formalin killed *Pasteurella tularensis*, strain No. 38 Francis.
This strain was originally supplied to the Ontario Public
Health Laboratory by Dr. E. Francis of Washington.   As it
has not been used for animal inoculation, it has lost its
virulence, but is perfectly effective for agglutination tests.
Any appreciable titre of *P. tularensis* agglutinins in the serum
of an animal that is not actually sick with tularemia means
that the animal has had the disease and recovered.   The
findings of Green and Shillinger (1932) showed that the
agglutinins in the blood of a hare infected in summer disappeared before a winter had passed, which is in sharp con-
trast to the striking persistence of the agglutinins in humans
(Francis, 1928; Thiotta, 1931).   Of twenty-nine sera tested,
four gave a positive agglutination (see table 15).   Two of
these were from Arden in Frontenac county, one from Bradford in Simcoe county, and one from Smoky Falls on the
Mattagami river, a wide distribution.   That from Bradford
had a very high titre but showed no lesions except a small
mass of pus under the skin of the chest.   Only *Staphylococcus
aureus* was cultured from this, and a guinea pig injected with
the pus and surrounding tissue did not become sick.

A Cottontail Rabbit Epidemic

An epidemic among cottontail rabbits (*Sylvilagus floridanus*) at Abbey Dawn near Kingston has been observed by
the writer in March, 1934; during one day's searching eleven
dead rabbits were found, and others had been seen during the
preceding three weeks by Mr. W. H. Robb.   In all specimens
that were in suitable condition it was found that the lungs
were pneumonic and the immediate cause of death appeared

to have been suffocation by pleural hemorrhage. The lungs of one rabbit that had been found dead and two that were shot, proved sterile on blood agar, a fourth (found dead) yielded *Staphylococcus albus* and a fifth (found dead) harboured colon bacteria, *Escherichia* sp. Blood sera of six rabbits, five of them freshly shot, one found dead, were tested for *Pasteurella tularensis* agglutinins with uniformly negative results. The causative agent was not discovered, but it was practically proven that the epidemic was not due to tularemia.

TABLE 15.—Positive agglutination tests on varying hare sera with *Pasteurella tularensis* antigen, macroscopic tests.

Hare number	Locality	Date	Reading of test
65	Arden, Frontenac county	Nov. 9, 1932	Agglutination complete to dilution of 1:80
135	Arden, Frontenac county	Jan. 9, 1933	Agglutination complete to 1:160
98	Bradford, Simcoe county	Nov. 2, 1933	Agglutination complete to 1:5120 partial at 1:10, 240; inhibition from 1:80 down
162	Smoky falls, Mattagami river, Ont.	July 4, 1935	Slight agglutination in 1:25 and 1:50 dilutions

INTERNAL PARASITES

The general revisions or studies of helminths of rabbits contain few references to those of the varying hare. The only reference in Stiles (1896) is to the effect that Curtice (1892) reported *Taenia pectinata* (probably *Cittotaenia variabilis*). Hall's (1916) revision of the nematodes contained no record for the hare under consideration; likewise the revision of nematode parasites of vertebrates by Yorke and Maplestone (1926) apparently lacks records for this hare. Boughton (1932) made a survey of the helminth parasites of the varying hare in Manitoba, describing two new species. This paper will be mentioned only when making a definite comment.

Sample specimens of the following parasites were identified

by the Bureau of Biological Survey, Washington, except *Cittotaenia pectinata*, *Multiceps serialis*, and *Synthetocaulus leporis*. Localities where specimens were found are listed and are in Ontario unless otherwise stated.

PROTOZOA: No protozoa have been found in the blood smears. No case of serious infection with coccidia was encountered, but their presence was noted occasionally.

TREMATODES: None found.

CESTODES, TAPEWORMS: (1) *Cittotaenia pectinata*. Adult tapeworms were scarce. References, Boughton (1932). Localities: Frank's bay, Markdale, Plevna.

(2) *Taenia pisiformis*. Larval cysticerci or bladders occurred commonly in small numbers in about 40 per cent. of the hares examined, which is somewhat more frequent than found by Boughton (1932), *e.g.*, 14 per cent. Other references: Harkin (1927) in British Columbia, Law (1933) in Ontario. Localities: Arden, Biggar lake in Algonquin Park, lake Couchiching, Coboconk, Fergus, Frank's bay, Huntsville, Hogarth, Kapuskasing, Kingston, Metagama, Minesing, Oba, Plevna, Pottageville, Smoky Falls on the Mattagami river, Stratford, Sundridge, Timagami, Woodbridge, Woodville, and Paswegin, Saskatchewan.

(3) *Multiceps serialis*. This larval form makes the large multiple cysts that are so prominent as to be frequently reported by our correspondents. There is a suggestion that it is more common in western Canada than in the east (see table 16). The bladders can interfere with the use of limbs or the functioning of other organs and may cause death indirectly. In Ontario it has been found in less than 10 per cent. of the specimens. References: Boughton (1932), Harkin (1927), Seton (1909, 1928). Localities: Hogarth, lake Couchiching, Plevna, Woodville, and Paswegin, Saskatchewan.

NEMATODES, ROUNDWORMS: (1) *Trichuris leporis*. In large intestine. Reference: Boughton (1932). Localities: Bradford, Frank's bay, Plevna, Oba.

(2) *Nematodirus triangularis*. In duodenum, Boughton (1932). Locality: Frank's bay.

(3) *Protostrongylus leporis*. In duodenum. Frank's bay.

(4) *Trichostrongylus calcaratus.* In duodenum and cae-
cum. Frank's bay.

(5) *Obeliscoides cuniculi.* Rabbit Stomach Worm.
Numbers up to one or rarely two hundred of these blood-
sucking worms were found in the stomachs of nearly all the
hares autopsied, and may be considered "normal". Its life
history has been described by Alicata (1932). Infection is
by feeding. The larvae are of three stages, the third in-
fective, and reached in about six days. These worms were
the cause of the only epidemic among hares studied in the
field. In the summer of 1935 at Smoky Falls on the Mattagami river recently captured hares began to die, and they
had not been mistreated. Of seven that died six had the
stomach full of these worms in yellowish, thick fluid. The
hares did not eat for a day before dying in convulsions. The
liver was dark red and congested.

Negative results were obtained from bacterial cultures
from the liver, blood, and brain, as they were sterile. Guinea
pigs inoculated with materials from the dead hares remained
well. Brain tissue rubbed on the scarified cornea of a varying
hare caused no harm. Blood serum contained no agglutinins
for *P. tularense.* Blood smears were negative for protozoa.
These methods should have detected infection with bacteria,
filtrable virus, or protozoa. It seems definite then that the
stomach worms caused these deaths.

Seton (1909, 1928) has recorded deaths from the same
parasites.

Localities: Arden, Bradford, Biggar lake in Algonquin
Park, Coboconk, lake Couchiching, Fergus, Frank's bay,
Huntsville, Kapuskasing, Kingston, Minesing, Oba, Pottage-
ville, Plevna, Savanne, Smoky Falls on the Mattagami river,
Timagami, Woodbridge, Woodville, and Lake St. Martin
Indian Reserve in Manitoba.

(6) *Passalurus ambiguus.* In large intestine. Reference:
Boughton (1932) as *P. nonannulatus.* Locality: Frank's bay.

(7) *Wellcomia evoluta.* Found by Dr. Seymour Hadwen
in a hare from Cartier, Ontario.

(8) *Synthetocaulus leporis.* Lung worms were only occasionally found in the trachea and lungs, and in small numbers.

Reference: Boughton (1932). Localities: Frank's bay, Metagama, Plevna.

(9) *Dirofilaria scapiceps.* Only one hare was found infected with filarian worms. It was taken near Plevna, Frontenac county, and had 5 filarian nematodes in the connective tissue between the peronaei tendons and the extensor tendons of the right hind leg. They had interrupted the anterior tibial artery and vein, but there was little free blood. The two groups of tendons were eaten into, the peronaei being particularly weakened. Microfilaria, the larval forms, were seen in blood smears.

OTHER WORMS: Reports of worms along the back have come from three correspondents, but as these have not been found in the autopsies, no identification was possible. They might be bot larvae, or nematodes. Mr. L. O. Foster of Shining Tree, Ont., described it as "a whitish grub, ½ inch long by 1/8 inch through, lying in the centre along the spine under the pelt". Mr. R. K. Wilson reported: "Jan. 1, 1934, small white worm noted along back bone, inside skin . . . size of match head and stretches lengthwise when not bothered."

EXTERNAL PARASITES

A briefly annotated list of the ectoparasites is presented here. Species which were identified by the Taxonomic Division of the Bureau of Biological Survey at Washington are marked with "(W)". All were from the varying hare (*Lepus americanus*) in Ontario, unless otherwise stated.

(a) *Acarina—Parasitic Mites*

(1) *Haemogamasus alaskensis* Ewing. (W). This gamasid mite of the family Parasitidae, a true blood sucker (Ewing, 1929), was found on a hare from near Batchawana bay, Algoma.

(2) *Trombicula microti* Ewing. (W). A chigger, in the family of harvest mites or Trombidiidae, of which the larvae only are parasitic. Localities: near Batchawana bay, Algoma, and at Biggar lake, in Algonquin Park.

(3) *Cheletiella parasitivorax* Meg. (W). Localities: Frank's bay on lake Nipissing and Smoky Falls on the Mattagami river.

(4) Adult beetle mites of the family Nothridae were found on a hare at Smoky Falls.

(b) *Acarina—Ixodoidea—Ticks*

(5) *Haemaphysalis leporis-palustris* Packard (1869). (W). This tick, the usual one on rabbits and hares of many species, has been recorded from the varying hare by Nuttall and Warburton (1915) at Aweme, Man., and by Hewitt (1915). It has been referred to loosely as "ticks" without specific identification by Seton (1909, 1928), Snyder and Baillie (1923), and others.

In the present study the specimens identified at Washington came from Smoky Falls on the Mattagami river, as did some that Mr. R. V. Whelan sent to Dr. Nathan Banks of the Museum of Comparative Zoology at Cambridge, Massachussetts; some identified by Dr. R. Matheson of Cornell University were from Hornings Mills and Plevna. The writer has identified about a thousand adult ticks off varying hares in Ontario, and handled many "nymphs" and "larvae", without finding any other species.

The total numbers of ticks on hares taken at the four main field study stations, and their relations to the hare cycle of abundance are discussed in the sections on correlations.

The species has been collected from hares in widely separated parts of Ontario, as shown by the following list of additional localities: Arden, Batchawana, Biggar lake in Algonquin Park, Dacre, Frank's bay on lake Nipissing, Mansfield, and Sundridge.

(6) *Dermacentor venustus*. Thomas and Cahn (1932) stated that this tick "has spread in northern Minnesota to . . . snowshoe rabbit".

(c) *Insects—Siphonaptera—Fleas*

(7) *Monopsyllus vison* (Bak.). (W). Fleas were rarely encountered, occurring on only two hares,

one from near Batchawana bay in Algoma, and the other from Kapuskasing, Ontario.

(d) *Insects—Diptera—Flies*

(8) *Aedes* sp. (W). A female mosquito collected off a hare at Smoky Falls on the Mattagami river. A hare kept in a small cage on the ground several times became frantic due to the multitude of mosquitoes that gathered in the cage.

(9) *Simulium venustum.* (W). Found feeding on a hare at Smoky Falls.

(10) *Cuterebra* sp. (W). Although no bots were found in varying hares by the writer, Mr. A. H. Burk gave an account of an infection of epidemic proportions, which has been quoted in the section on correlations of the hare cycle with diseases. A third instar larva was collected from a cotton-tail rabbit (*Sylvilagus floridanus*) at Turkey point in Norfolk county.

CORRELATION OF THE CYCLE OF ABUNDANCE WITH OTHER
PHENOMENA

It is now proposed to deal with the evidences as to possible causes and effects of the cycle.

The decrease in numbers of hares in a locality is due to wholesale dying off rather than to emigration away from that area or excessive predation by animals that eat hares. If the hares emigrated, it would have to be northward in Ontario, since they are abundant later to the north than south. Then where do the hares go that compose the abundance in the north? The great magnitude and suddenness of the decrease have been pointed out above. If mass movements of that size took place they would have been noticed, but there is no such record. Foxes, lynxes, and other predators do not increase to such numbers as to be able, within one year, to dispose of nine-tenths of the several thousand hares to the square mile that constitute abundance; moreover, hares disappear periodically in large areas where lynxes have been all but exterminated; further, the whole population of predators has been greatly modified by man, yet the hare cycle still continues. Several observers have actually found many dead hares in the woods at times when hares were decreasing in abundance (Seton, 1909, 1928). Reports from questionnaire correspondents of similar occurrences are quoted below. In 1926 in Newfoundland, Mr. F. R. Hayward "saw many sick rabbits, too dopey to get out of the way of anything; many were killed by trains. Rabbits were more abundant than I ever saw them before or since. Counted 19 dead rabbits in one clearing." Dr. C. D. Howe in 1909 in Nova Scotia saw sick and dead rabbits frequently. In 1910 he saw many such but not so many as in 1909. In the country surrounding Montreal Mr. C. Perrault "started to find dead hares in the fall of 1931 and many more dead ones this fall of 1932". In May and June, 1923, Mr. H. C. Haskins "found hundreds of dead ones" in Nipissing and Timiskaming districts, Ontario. In the same year, 1923, Mr. H. Elkington records the following for the north parts of Frontenac county and the combined county of Lennox and Addington: "Rabbits were sitting around and could be picked up, they were so sick. Most of them died. They seemed to be polluted with ticks and stunk before they were dead." Speaking of the same time and place, Mr. P. J. Wensley said: "The snowshoe rabbits died off in large numbers. A stench was noticed all along the road. Many unused telephone pole holes contained one or two dead rabbits." Mr. W. D. Gouldie of Dwight, Muskoka district, Ontario, saw dead hares in April, May, and June, 1934. Dead hares were found lying on snow in woods in vicinity of Armstrong, Thunder bay district, Ontario, in the spring of 1934 by Mr. F. Belmore. Mr. Sam Waller reported many frozen and starved during November, December, and January of the winter 1933-4 on the Lake St. Martin Reserve, Manitoba. In Saskatchewan in January, 1933, Mr. A. E. Etter, game commissioner, told us "Reports being received now to the effect that hares are being found dead in large numbers, and in almost every locality, in the southern half of the province". During the late winter and early spring of 1934 in the Fort à La Corne Forest and Game Preserve lying north of the area mentioned above, Mr. J. M. Brown reported that "varying hares were found dead by the

thousands all through the bush". In the vicinity of Red
Deer, Alberta, Miss M. P. Cole said that "in 1899 rabbits
were sick, dying with sores on throats. People unfortunate
enough to eat the meat of sick animals were very ill. I have
seen remains of hundreds of rabbits on one half section."
Mr. Axsel Smith, game inspector, of Athabaska, Alberta,
"found these animals dead in large numbers on all trails in
the year 1925", and in January, 1934, reported: "Now dying
off in large numbers, the woodlands are covered with dead
hares. Have inspected many dead hares and found them
all diseased, that is, poor, and of no weight at all. I have
observed the young hares dying before changing colour last
Fall, and seen them die on the roads of weakness." Mr. C. F.
Archer reported many dead "with white water blisters under
the skin", in the vicinity of Kootenay lake, British Columbia,
in the fall of 1914 and during the next year. A game super-
visor, Mr. C. F. Kearns, reported as follows of the south-
eastern part of British Columbia: "In the summer and fall of
1925 vast numbers of snowshoe rabbits were observed to be
sick and dying, and for the succeeding years they were comparatively scarce." An interesting account comes from Mr.
Hamilton M. Laing for the region near Mount Logan, eighty
miles from McCarthy, Alaska: "The rabbits were dying off
in the McCarthy region when I went in in the spring (1925).
Most had died in the white winter coat."

Correlation with Rate of Reproduction

It has been stated that varying hares have usually three
or four young in a litter, but that during the times of increasing abundance they have as many as eight or ten. This has
been said by numerous writers, including MacFarlane (1905),
Preble (1908), Seton (1909, 1911, and 1928), Hewitt (1921),
and Elton (1924). Actual evidence in support of the claim
is not presented, except by Preble, who said, "in the few
instances where I ascertained the number it seldom exceeded
two", in the Athabaska-Mackenzie region in 1904 when hares
were decreasing in numbers.

Mr. R. V. Whelan examined the hares shot by the residents of Smoky Falls on the Mattagami river between May 17

and June 15, 1933, and kindly contributed his findings. Of the females, three carried one young, two carried two young, twelve carried three, and one carried five; the average being 2.7 per litter. Two-thirds of the specimens carried three young. At that time hares had just reached the condition of abundance. In 1935 during the corresponding time of year, of eight females, one had one young, six had two, and one had three; the average being two per litter. Three-quarters of the litters contained two young. The hares were still abundant but had started their decrease in numbers. These averages indicate a significant difference in the rate of reproduction in those two years at that place, but whether it was biologically related to the population cycle remains an open question.

At Frank's bay the average numbers of young per litter (table 9a) were as follows: in 1932, based on five litters, 3.4 young; in 1933 the year of peak abundance, based on three litters, 3.3 young; in 1934, based on three litters, 2.7 young. These figures are less reliable than the Smoky Falls set, being based on fewer records; but they suggest a similar change in the rate of reproduction. No records are available for the years in which hares were increasing in numbers.

It seems impossible that changes in the reproductive rate can account for decrease in numbers such as occur at times of periodic disappearance. The magnitude of the decrease is too great within one season. For instance, at Buckshot lake in 1932 the hares decreased by September to about one-fourteenth of their numbers in July, and local residents said they had been more numerous in the spring than they were in July. This means that adult hares must have died in great numbers, even if it were assumed that no young at all had been produced that season.

Correlation with Diseases and Parasites

Disease as a factor in the decrease of animal populations has long been recognized, for Darwin (1859) wrote: "When a species, owing to highly favourable circumstances, increases inordinately in a small tract, epidemics—at least this seems generally to occur with our game animals—often ensue; and here we have a limiting check independent of the struggle for life.   But even some of these so-called epidemics appear to be due to parasitic worms, which have from some cause, possibly in part through facility of diffusion amongst the crowded animals, been disproportionately favoured; and here comes in a sort of struggle between the parasite and its prey." The different classes of pathogenic organisms will now be discussed in turn.

(a) *Miscellaneous diseased conditions*

Certain features of disease presumed to be responsible for the death of hares have been mentioned by correspondents. A lethargic, "dopey" condition, in which the hares may be almost touched before they move sluggishly away, was reported relatively more often for the period of decrease between 1919 and 1928 than for the recent period of decrease in the 1930's (see table 16).   This condition has not been identified with any particular disease, since during the last few hours of life a hare sick from almost any disease is usually sluggish.   Carcases of dead hares were also found more often in the 1920's than in the recent period of decrease, which suggests that the previous dying off was due to a more severe and noticeable epidemic than was the recent decrease. Corporal E. S. Covell of the Royal Canadian Mounted Police stationed at Moose Factory, Ont., said that the epidemic of 1934 "did not appear to spread so disastrously as in other years".   From the experience gained in the present study, it is known that reports of ulcers, sores, and pus usually referred to cases of infection with the pyogenic germs, *Staphylococcus aureus*, and they occurred as often during the decrease in the 1920's as during that of the 1930's.   Symptoms of watery eyes and discharge from the nose, similar to those of the disease known as snuffles, were noted four times, that is, in only 1 per cent. of the reports, during the decrease of the 1930's.   Several observers in describing a sick animal suggested a septicemia, and others noted diseased livers.   Practically all the reports of disease referred to times when the hares were decreasing.

TABLE 16.—Reports of disease from correspondents.

Section (*a*).—During the decrease in abundance of hares between years 1919 and 1928.

	Newfoundland	Nova Scotia	New Brunswick	Quebec	Ontario	Manitoba	Saskatchewan	Alberta	British Columbia	N. W. Territories	Yukon	Alaska	Totals	Per cent.
Lethargic condition.	1	..	..	..	11	..	..	1	..	..	..	1	14	18
"Sick".	1	..	..	..	..	..	..	..	1	..	..	1	3	4
Dead hares found. . .	1	..	..	..	11	..	1	2	6	..	..	1	22	29
Ticks, suggested as cause of death, or at least as a serious condition. . . .	..	..	..	..	5	..	1	..	..	..	..	1	7	9
Ticks, occurrence alone mentioned. .	..	..	..	..	(4)	..	..	..	..	..	..	..	..	..
Fleas (this and previous class not included in totals). .	..	..	..	..	..	..	..	..	..	..	..	..	..	..
Sores, ulcers, pus. . .	..	..	..	..	4	..	1	..	..	..	..	..	5	7
Liver, diseased.	..	..	..	..	..	..	..	1	..	..	..	..	1	1
Tapeworms, adult. .	..	..	..	..	1	1	..	..	..	..	..	..	2	3
Tapeworm, larval cysts, *Cittotaenia ctenoides*	..	..	..	..	1	..	..	..	1	..	..	..	2	3
Tapeworm, larval cysts, *Multiceps serialis* "water blisters".	..	..	..	..	11	..	2	2	1	..	..	..	16	21
Lumps, possibly tapeworms larvae or pus.	..	..	..	..	1	..	..	..	..	..	..	..	1	1
Miscellaneous.	..	..	..	..	2	..	..	1	..	..	..	..	3	4
Totals.	3	0	0	0	47	1	5	7	9	0	0	4	76	100

(b) Bacteria

MacFarlane (1905) spoke of the "virus of the disease which periodically affects the head and throat, and carries off many thousands of the American Hares".

Section (b).—During the decrease in abundance of the varying hare between years 1929 and 1936.

	Newfoundland	Nova Scotia	New Brunswick	Quebec	Ontario	Manitoba	Saskatchewan	Alberta	British Columbia	N.W. Territories	Yukon	Alaska	Totals	Per cent.
Lethargic condition.	..	..	..	..	22	1	3	3	..	..	..	..	29	7
"Sick"	..	..	1	..	19	4	6	2	1	..	..	..	33	8
Dead hares found	1	1	1	2	58	17	11	3	4	2	..	..	100	24
Ticks, suggested as cause of death, or at least as a serious condition	..	..	2	..	41	2	3	1	1	..	..	..	50	12
Ticks, occurrence only	(1)	(3)	(7)	(2)	(126)	(16)	(15)	(6)	(5)	..	..	..	..	..
Fleas (this and previous class not included in totals)	..	..	(1)	..	(5)	(2)	(2)	..	(1)	(2)	..	..	..	..
Cuterebra, bots	..	..	..	..	1	..	..	..	..	..	..	..	1	⅛
"Skin disease"	..	..	..	..	8	1	1	..	..	..	..	..	10	2
Sores, ulcers, pus	..	..	..	1	21	1	4	3	..	..	..	..	30	7
Watery eyes, discharge from nose.	..	..	..	..	4	..	..	..	..	..	..	..	4	1
Liver diseased	..	..	..	..	2	1	..	3	2	..	..	..	8	2
Tapeworms, adult form	..	1	..	..	3	..	1	..	1	..	..	..	6	1
Tapeworm, larval cysts, *Cittotaenia ctenoides*	..	1	..	..	9	..	2	2	2	..	..	..	16	4
Tapeworm, larval cysts, *Multiceps serialis* "water blisters"	..	..	..	2	45	10	18	6	2	..	..	..	83	20
Worms along back	..	..	..	..	3	..	..	..	..	..	..	..	3	1
Other worms	..	..	..	..	6	..	..	..	..	..	..	..	6	1
Lumps, possibly tapeworm larvae, or pus	..	..	1	..	11	3	2	4	1	..	..	..	22	5
Starvation	..	..	..	..	..	1	..	..	1	..	..	..	2	½
Frozen	..	..	..	..	..	1	1	..	..	1	..	..	3	1
Miscellaneous	..	..	..	..	10	1	..	5	1	..	..	..	17	4
Totals	1	3	5	5	263	43	52	32	16	3	0	0	423	100

Along the Hay river and Lower Athabaska river in the summer of 1903, Preble (1908) reported that a large proportion of the rabbits that were shot had accumulations of pus beneath the skin of the neck. In January and February, 1904, "many thousands of rabbits perished from disease. . . . The throat and lungs were much inflamed".

An epidemic occurred among the varying hares and other hares and rabbits on Seton's estate in Connecticut in late summer, 1903. Those found dead "appeared to have their throats cut", and the hare that was autopsied by a medical doctor had died of infection with the common germ of suppurative processes, *Staphylococcus aureus* (Seton, 1909, 1928).

In 1920, in Utah, Francis (1925) found seventeen jack rabbits infected with tularemia out of a total of five hundred and fifty-six shot in a stretch of country sixty miles in length. This is as great a frequency, he said, as would be expected in a population which was being destroyed by an epizootic disease.

Green and Shillinger (1934) found that in the lake Alexander area, Minnesota,-in 1933, "the occurrence of tularemia in the ticks found upon the snowshoe rabbits rose to 0.24 per cent. during April and May. With this marked rise in tularemia in ticks, the disease was found to be present in epidemic proportions among the snowshoe rabbits, but no appreciable mortality developed." They reported (Green and Shillinger, 1935) a decrease in the hare population between the spring of 1933 and that of 1934. Tularemia was found to have increased among both hares and ticks. "It is only reasonable to ascribe to the observed spread of tularemia an important role in these losses", they wrote.

In the above area they demonstrated (1932) that cottontail rabbits (*Sylvilagus floridanus*) decreased in numbers in the summer of 1932, and that tularemia occurred as a constant infection in sick and dead animals.

A destructive epidemic of tularemia affecting both sheep and jack rabbits occurred in the north-western states in the spring of 1934 (Anonymous, 1935), but none of them in epidemic form correlated with a decrease in the population of hares. Although a hundred and sixty specimens have been examined for disease, during the present investigation no unusual incidence of infection with filtrable virus or bacteria has been demonstrated (see the report on bacteriology). Evidences of various diseases were found, such as staphylococcosis and tularemia.

(c) *Protozoa*

Ritchie (1926) told of epidemics among alpine hares in England and Scotland, one of which was proved to be due to coccidiosis. The present study has not detected any epidemic caused by coccidia or other protozoan parasites.

(d) *Worms*

Two snowshoe hares from Maine were sent in October, 1904, to Seton (1909, 1928). They died in a short time and were examined by Dr. W. Reid Blair, of the New York Zoological Park, who reported they had been killed by roundworms, chiefly stomach worms, *Strongylus strigosus* (now called *Obeliscoides cuniculi*): "The parasites were present in great numbers, many hundreds in each animal." Dr. W. T. Hornaday, who took part in the investigation, was of the opinion that these accounted for the "periodical seven-year plague among the northern Varying-hares". Collett (1911-2) stated that alpine hares were subject to epidemics in Norway in certain years, causing disease of the lung and associated with strongylid worms. For the year 1933 in Ontario, Law said (1933): "There is evidence from examinations made this year that rabbits (referred to varying hare) are reaching a high peak of parasitism." He had just mentioned tapeworms.

The questionnaire correspondents (table 16) have mentioned the "large watery blisters with small white particles in them", which are the larval stages of the tapeworm (*Multiceps serialis*). Many of them blamed this for causing the decrease in abundance of the hares. This condition is so prominent that it is noticed by even the casual observer, and it is therefore not surprising that it was mentioned in 20 per cent. of the reports received during each of the last two periods of decrease. The bladders of the larval tapeworm (*Cittotaenia ctenoides*), in the body cavity and under the serous coats of intestinal organs, adult tapeworms within the intestines, and other worms were mentioned less frequently.

The one worm parasite whose occurrence has been correlated with the cycle in the present investigation is the stomach worm (*Obeliscoides cuniculi*) (see the report on internal parasites). A moderate number of these blood-sucking worms have been found in the stomachs of nearly all the hares autopsied, and may be considered "normal". However this stomach worm was the cause of death of six out of seven well-cared-for hares that died "natural deaths" within several days of capture at Smoky Falls on the Mattagami river in the summer of 1935. As has been shown, the hares were beginning to experience a major decrease in numbers late that summer. The evidence supports the view that the decrease at that time and place was caused by stomach worms. Isolated cases of serious infestation with these worms were found at some other places, but not in epidemic form.

(e) *External parasites*

Seton (1909, 1928) mentioned a number of cases of hares carrying many ticks on their heads and ears at times of peak abundance and decrease. In 1922 near Brent, Ontario, specimens examined by Snyder and Baillie (1923) "were infested with wood ticks, especially on the ears and head". Green and Shillinger (1935) showed that "The rabbit tick population appears to have passed a peak in 1933 along with the hare population, but the two occurrences are probably not directly related", in Minnesota. Ticks are known to carry diseases, such as tularemia (Parker and Spencer, 1924), tick paralysis (Hadwen, 1913), anaplasmosis (Rees, 1934) and East Coast fever (Cowdry and Ham, 1932), and their abundance, therefore, might be an important factor in the attainment of epidemic proportions by a disease. Green and Shillinger (1935) found that tularemia was not present among snowshoe hares in winter when ticks were absent; and the disease appeared again with the ticks in the spring.

About 10 per cent. of the reports of disease from questionnaire correspondents suggested ticks as the cause of death or as producing at least a serious condition.

Mr. A. H. Burk has described an epidemic which appeared to be caused by the larvae of the Oestrid fly (*Cuterebra* sp.). The region was that between Sault Ste. Marie, Blind river, and the south part of the Mississagi Forest Reserve, and the year, 1915. "Rabbits, previously plentiful, were dying in large numbers. In one day's walk in the bush you could count up to sixty dead rabbits which had died within a few days, and many live rabbits in such a state that they were unable to run and could easily be picked up. In all cases dead or dying rabbits were infested with a parasite, the species I do not know. The parasite seemed to be a larva or pupa very much like the pupa of a large moth. They were approximately one inch long by one quarter inch in diameter, and they formed large lumps directly under the skin of the rabbit, usually around neck and hind legs. A small hole through the skin of the attacked rabbit was at one end of the larva and this perforation was always very festered. I might add that at this time I had about ten domesticated Belgian hares, and they were all attacked by the same parasite, ending in the death of all but two. These two were saved by the fact that I cut three parasites out of each of them." It is surprising that these bot flies should cause death but the account is convincing; it is not impossible that the deaths were due to some other cause.

The correlation noted by Green and Shillinger (1935) of many ticks per hare at a time of abundance of hares has appeared in this investigation. The records are not sufficient to indicate that this always accompanies a peak of abundance. At Frank's bay the numbers of ticks per hare in the summers of 1932, 1933, and 1934, averaged from nine, six, and seven hares respectively, were 120, 340, and 510. Hares reached the peak in 1933 there and were decreasing but still quite abundant in 1934. At Buckshot lake for the summer of 1932 the average from ten hares was a thousand ticks per hare; at this time the hares were decreasing but very abundant. In the next summer the average from two hares was 2,700 ticks each. The average number of ticks from ten hares at Smoky Falls on the Mattagami river in 1935 was 580. This increase of ticks per hare is understandable, because the "larval" ticks, and nymphs and adults after moulting, would have a better chance of getting onto a hare when the latter were abundant. ·Fewer individual ticks would be wasted, and more females would mature to lay eggs.

In Algonquin Park in the old forest of big trees described earlier, ticks did not become abundant; the most taken off one hare was six. The hares died off there, which shows that the abundance of ticks is not a necessary factor for the decrease of hares; it took place there without them.

Correlation with Food

Food shortage has been suggested as a cause of decrease in animal populations. It is therefore worth considering whether hares may become more numerous than the available food will support, with the result that the usual decrease would take place by starvation directly and indirectly by failure to withstand bad weather in an emaciated condition. This situation was hinted at by Seton (1911) who said: "Finally, they are so extraordinarily superabundant that they threaten their own food supply . . . everywhere were acres of saplings barked at the snow-line." Mr. A. C. Cutten of Iroquois Falls, Ontario, wrote in answer to a questionnaire: "As the snow melted before they died off the last time they attacked the young banksian pines, then about six to eight feet high after the fire of 1916. They ate all the green and the outer bark so that the stands of the trees showed a band of cream colour visible for a long distance. Normally they do not eat jack pine but they are now attacking them, but not eating the bark. The writer believes that there is some relation between the attack on the jack pine and the plague." Mr. J. A. Brodie of the Ontario Forestry Branch observed jack pines and spruces that had been damaged by hares in the vicinity of Loch Lomond in Thunder bay district, Ontario, and pointed out that it was possible to determine the date of the last abundance of the hares by counting the rings of growth on the trees subsequent to the inquiry.

The specimens collected at times when the hares were decreasing in numbers, as at Buckshot lake in 1932 and at Smoky Falls in 1935, had ample food in their stomachs. They were not fat, but did not appear to be starving. Any obscure dietary deficiency would not have been detected. At Buckshot lake no evidence of depletion of their food supply was noticed. On plot 21, at Smoky Falls the hares had eaten the bark off a large number of small poplar trees during the winter of 1934-5. Only a few trees had been barked the year before, 1933-4, when the hares had been at least as abundant. The hares would appear to have had enough food in the winter of 1934-5, but an equal amount would not have been available for the same effort the next winter. If the hares had not been killed off by disease during the summer and had increased in numbers, they would have felt the pinch of starvation in the winter of 1935-6. The forest was not producing bark at the rate it was consumed in 1934-5, but considered over the ten year cycle sufficient bark was certainly produced to stand the one year's heavy cropping. Some old forests might lack sufficient small trees and saplings to feed a large population of hares.

Correlation with Habitat

The lack of food suggested above might be the check that moderates the cycle in old forests. In Algonquin Park the comparatively low peak of numbers and the relatively high values for the years 1934 and 1935, when the hares had changed from what was locally abundant to what was locally scarce, in comparison with the population figures for other localities, suggest the cycle was reduced in violence in this heavily forested region. The park ranger for that district, Mr. R. Edwards, said: "Rabbits are never very abundant in this wooded country; they do not fluctuate as drastically as in brulé country."

A second correlation of habitat with the cycle has been elaborated in the section on life history, where it was shown that in times of scarcity all the hares in the country are in the swamps, the "typical" habitat, but in times of abundance they spread into all possible areas.

Correlation with Cycles of Abundance of Other Animals

Hewitt (1921) and Elton (1924) have shown that lynx, fox, marten, fisher, and mink are subject to cycles of about the same length as that of the varying hare.

The fur returns for lynx (*Lynx canadensis*), of the Hudson's Bay Company, are shown in figure 16 (taken from MacLulich (1936c)). The year 1789 is taken as a peak for calculations, as it is central in a group of three low modes. It is a generally recognized fact, borne out by the fur records, that lynx abundance is governed by that of rabbits, their staple food. Quoting MacFarlane (1905, 1908): "The yearly catch of lynxes rapidly diminishes in volume as soon as the rabbits become scarce and when the latter are comparatively rare a large proportion of the great but now dwindling crowd of lynxes suffer privation, and some actually starve to death." Seton (1925) narrates that during the winter of 1906-7 in the Mackenzie river valley, "I met with a dozen lynxes that were dying of starvation—mere walking skeletons—and in the silent woods found a dozen shrivelled corpses". None of the lynxes he examined had any rabbit in its stomach nor much of anything else. Dearborn (1932) showed that the varying hare comprised 89 per cent. of the food of the bay lynx (*Lynx rufus*) in Michigan. The decrease in abundance of the rabbits is abrupt and rapid; they "almost disappear within one or two years" (MacLulich, 1935). Several of the rabbit peaks, suggested by the early inadequate fur returns, are in close agreement with the lynx peaks. Since their peaks in 1856 the average lengths of the cycles have been 9.8 years for the rabbits and 9.9 years for the lynx. Although the accounts (Seton, 1911; Preble, 1908) of the lynx abundance lagging a year after the rabbit peaks are convincing, yet the averaged results given by the Company fur returns fail to show this. The lags of the lynx peaks after the rabbit peaks were: 0, +2, 0, −1, −1, 0, −1, +1, and averaged zero. The strength of association between rabbit and lynx abundance may be measured by the coefficient of correlation. A coefficient of "one" means perfect correlation and "zero" means complete lack of correlation. If the coefficient is less than its probable error, there is no correlation and if it is greater than six times its probable error there is definite correlation (following Tuttle and Satterly, 1925). A positive coefficient means the two variables increase together; a negative one means one increases while the other decreases. The coefficient of correlation will be used hereafter in this

paper without further explanation. For this calculation the
abundance of the rabbit and that of the lynx were read off
the graphs of figures 16 and 17 as scarce, intermediate, or
abundant, for each year from 1847 to 1934 inclusive.   This
simple treatment was necessary, due to the varied nature of
the scales of different sections of the graphs; the coefficient
calculated from it compares the synchronism of the cycles.
The coefficient of correlation of the rabbit and lynx abundance is +0.55 with a probable error of only one-eleventh
of that value, *e.g.*, ±0.05, which shows definite correlation.
Therefore there is good ground for believing the decreases in
numbers of lynxes are caused by starvation when the hares
disappear, or at least by inability to withstand adverse
circumstances and winter conditions on short rations.

The cycles of the red fox (*Vulpes fulva*) are not so pro-
nounced as those of the lynx, which "is probably accounted
for by the fact that, while the fox feeds upon the rabbit,
especially when the latter animal is abundant, it also feeds
largely on mice or voles, supplementing this diet with game-
birds of various kinds" (Hewitt, 1921).   In Michigan during
1930 and 1931 the food of foxes contained 47 per cent. by
frequency, or 81 per cent. by volume, of rabbit and hare
remains (Dearborn, 1932).   The years of fox abundance tend
to be a year or two later than the great rabbit years.

The peak years for marten (*Martes americana*) usually
preceded those for rabbits by a year or two.

The years of maximum abundance of fisher (*Martes
pennanti*) occurred close to, but usually later than, the rabbit
peaks.

The mink (*Mustela vison*) peaks occurred irregularly near
the hare peaks.   As hares form only a minor part of the food
of the mink, there seems no reason to expect a correlation in
their numbers, unless the two are affected more or less
independently by the same fundamental causes.

The coyote of the west or brush wolf of the east (*Canis
latrans*) has a diet similar in many respects to that of the fox,
except that it is more varied, including larger game.   Dearborn (1932), for instance, found that varying hares made up
83 per cent. by volume of the food.   It might be expected,
therefore, that there would be a correlation between the

numbers of coyotes and those of hares. Unfortunately the Hudson's Bay Company figures did not distinguish between brush wolves and timber wolves. The latter are much less dependent on hares than the former. This accounts for the absence of any very marked or regular periodicity in the fur returns. The numbers of coyotes taken in the west ("Canada Year Books", 1921 to 1935) had a peak in 1925-6 as listed, that is, 1924-5 would be the actual year of abundance. This is within a year of the peak of abundance of hares. The decrease of coyotes was slow unlike that of hares.

It is not yet sufficiently proven that these large animals are decreased in numbers by the decrease in abundance of hares. It will be possible to investigate this problem only when accurate local knowledge of their abundance and that of hares, mice, and other animals in the same area becomes available.

Some animals having no direct connection with hares, have ten or eleven year cycles of abundance, such as the Atlantic salmon (*Salmo salar*), as shown by Huntsman (1931); Pallas's Sandgrouse (*Syrraptes paradoxus*) of Asia (Elton, 1924); and the Siberian race of the Nutcracker (Simroth, 1908). The correspondence of these cycles must be due to simple coincidence or else to some world-wide controlling factor.

We have no evidence of the hare cycle applying to cotton-tail rabbits (*Sylvilagus floridanus*) or to European hares (*Lepus europaeus*) in America.

Correlation with the Weather

Seton (1911), Elton (1924), and others have looked to meteorological factors for some regulating principle governing all wild life cycles. No very positive suggestion has been made as to the effect of the individual factors of the weather. For instance, no rainfall figures have been compared with the animal cycles.

In an effort to determine if any correlation existed between precipitation and the decrease of hares in Ontario in the recent cycle, table 17 has been constructed from the monthly "Records of Meteorological Observations in Canada and

Newfoundland". This table indicates that, for this cycle, no such correlation existed.

An unusual disaster happened to the hare population of the low lands at the south end of James bay, Ontario, in 1934. Corporal E. S. Covell reported: "In May with the break-up of the river, water flooded most of this area and hundreds were found dead, hanging in the willows when the water went down."

Rowan (Elton, 1933a) suggested at the Matamek Conference that ultra-violet light might influence the cycle, owing

TABLE 17.—Precipitation in different regions of Ontario for the three months, May, June, and July, for various years concerned in the recent hare cycle.

Region	Precipitation					Years in which hares decreased
	1931	1932	1933	1934	1935	
Patricia	11.70	5.25	6.55	11.20	8.65	1933
Northern Ontario[1]	8.02	7.12	7.75	7.15	8.69	1934
Ottawa-Huron area[2]	8.06	6.62	7.49	6.25	8.09	1933
Frontenac area[3]	8.40	6.82	5.82	6.73	10.07	1932

[1]Average of Thunder bay, Algoma, Sudbury, and Timiskaming.
[2]Average of Nipissing, Manitoulin island, Upper Ottawa river, and Georgian bay counties.
[3]Average of Upper St. Lawrence and lower Ottawa rivers, lake Ontario counties, and east-central counties.

to experimental evidence suggesting a relation to disease resistance and to migration.

Correlation with the Abundance of Sunspots*

The most positive suggestion brought forward in regard to possible meteorological causes of the cycle is that of Elton (1924) that the hare cycle is related to the cycle of abundance of sunspots. The idea seems to be that the animal integrates the whole effect of small differences in the weather which are caused by the differences in solar radiation. This argument has been used and expanded by DeLury (1923, 1930, 1931) and Wing (1935). It has been shown by Thompson (1936)

*This section along with most of that on the fur returns has already been published (MacLulich, 1936c).

that such differences in weather are too small in North America to be measured by the meteorological stations. Furthermore, the present writer (1936c) has demonstrated that hare and lynx abundance are not correlated with sunspot abundance, but that the animal cycles are of different length and independent of the solar cycle. A copy of the pertinent part of this paper follows herewith.

In 1878 Swinton, in England, submitted inconclusive evidence that outbreaks of grasshoppers tend to occur at times of a minimum number of sunspots.

Ljungman (1880) compiled a record of the condition of the "genuine seaherring" fishery on the coast of Bohus län, finding good fisheries every hundred and eleven years. He claimed that there was a corresponding hundred and eleven year major period in sunspot abundance. In a footnote he admits, in regard to a correlation of governmental fisheries revenue, "during the present and the last century with Wolff's relative figures, I have not been able to find any very striking coincidence between the occurrence of solar spots and good fisheries".

Simroth (1908) pointed out that the invasions of Europe by the Siberian race of the Nutcracker (a bird) in 1844, 1864, 1885, 1896, and 1907 were correlated with the sunspot cycle.

Soon after Hewitt's book (1921) came out, DeLury wrote pointing out a correspondence between the fur records and the sunspot cycle. DeLury (1923) worked over some migration records from Montdidier in France from the year 1784 to 1869 and found the cuckoo arrived nine days later in years of sunspot maximum than in years with a minimum number of sunspots, the lark three days later, and the swallow only one day.

The present writer has just reinvestigated these records of bird arrivals, the migration dates being abstracted from DeLury's tables. Between the years 1813 and 1839 the cuckoo data are incomplete; there are figures for only eight years. This would suggest that the observers were not so thorough in their watching for the birds, and that the figures are probably not so reliable as for the other fifty-nine years. Using only the periods of regular observation, the difference in time of arrival is reduced from nine days to three and three-quarter days, while the correlation coefficients of sunspot number and date of migration of neither the cuckoo nor the lark are large enough to be significant.

The next year Elton (1924), in a comprehensive paper on cycles of lemmings, rabbits, and several other animals and birds, came to the conclusion that the periodic fluctuations in the number of these animals are due to the weather, which in turn is correlated with the sunspot cycle. He decided that the abundance of varying hares, as recorded by the Hudson's Bay Company fur returns, was related to the 11.2 year sunspot cycle. The discrepancies were explained by the changes in mean earth temperature, which showed cooling effects after major volcanic eruptions due to the dust emitted into the atmosphere. He went on to say: "If we allow for the irregularity in 1905, the rabbit period agrees very well with that of the sunspots; *i.e.*, omitting the one in 1905, the average period for rabbits between 1845 and 1914 is 11.5. But the 1914 one should have been in 1912, and this brings the average to 11.1, which is about that of the sunspots (=11.2)." He found no known cause for the short three and a half year cycles of lemmings and mice. The lack of complete simultaneity in animal fluctuations over the whole country was inferred to be due to the sunspot variation causing a cyclic shifting of the isoclimatic lines, with the result that rainfall increased at some places while it decreased at others. Julian Huxley (1927) in a popular article stated this more decidedly and also related the lemming cycles to a component of the sunspot cycle equal in length to one-third of the main cycle.

DeLury (1930) pointed out correlations of grasshopper and grouse maxima with sunspot minima; Criddle (1932) expanded the argument with much additional information on grasshoppers, but the correlation shown by the graphs is not convincing to the present writer.

From a study of outbreaks of certain forest insects, Eidmann (1931) has concluded that they are correlated with sunspot maxima.

Leopold in 1931 wrote as follows of grouse cycles: "There is no significant synchronism between Britain and America, the periods being different. This would seem to refute the theory that cycles are basically due to fluctuations in solar radiations or sunspots, unless such fluctuations operate through entirely different biological channels in America and Britain respectively."

At the Matamek Conference on Biological Cycles as reported by Huntington (1932) and Elton (1936), the sunspot cycle theory received considerable attention but not unanimous acceptance. DeLury suggested (to quote Elton, 1933b): "It is possible that the 9.7 year cycle in animal life is due to a climate cycle of short length caused by a blending of the 11.2 year sunspot cycle and the 8.85 lunar cycle in tides. Oceanic and inland reactions to the astronomic factors of opposite phase, were found, and illustrated by charts, resulting in half astronomical periods in some intermediate localities such as Anticosti."

Chapman (1933) said ". . . the whole matter of sunspot cycles and its effect upon biological phenomena rests upon a rather uncertain foundation."

In a note on sunspots and squirrel emigrations Anthony (1934) said: "A very interesting and, it might be said, troublesome coincidence exists between the periods of sunspot maxima and minima and the periods of plant growth . . . and the spectacular fluctuations of animal abundance. This coincidence is troublesome because it hints at too much to be overlooked, and yet the theories which attempt to show causal relations are by no means conclusive."

Wing (1934a, 1934b) showed a relation between migration times of various birds and solar cycles and developed the idea further with more data in 1935. Quoting (Wing, 1935): "The cycles are interconnected and composed of several elements but different species may respond to different components of solar activity, hence cycles may differ in amplitude, length, phase and latitude." He laid particular emphasis upon the Brückner cycle of 22.5 years, relating it to both migration dates and Hudson's Bay Company fur returns. In this connection it may be pointed out that the Brückner cycle is thirty-five years long rather than 22.5 years (Brückner, 1890, as summarized in Hanns, "Handbuch der Klimatologie", 1932). Leopold in a foreword to Wing's article wrote: "Apparently solar radiation is a common and cosmic denominator underlying the population behaviour of many species and conditioning their mechanism for population maintenance."

For comparison with the animal curves, the sunspot numbers have been shown on both figures 16 and 17. These are the "final relative sunspot-numbers, whole disc", annual means, compiled by the International Astronomical Union (Eidgenössische Sternwarte in Zürich, 1934).

Because of the close correspondence between rabbit and lynx cycles shown above, the conclusions that may be reached, regarding the relation or lack of it between lynx abundance and sunspot numbers, may be accepted as applying to rabbits also.

We may now proceed to compare the lynx abundance with sunspot numbers, for both of which data are available for a hundred and eighty-three consecutive years (see fig. 16). From the lynx peak in 1751 to that in 1925, a hundred and seventy-four years, there were eighteen cycles of lynx numbers but only fifteen and a half sunspot cycles. The average lengths of the cycles were 9.7 years for the lynx and 11.1 years for the sunspots. These facts in themselves mean the two cycles cannot be correlated, but are independent. The histories of the lynx, rabbit, and sunspot cycles are compared visually in figure 18. Each history is represented by a thick horizontal line along a scale of years, with the alternate cycles marked in solid black, and the others clear. It will be seen that the lynx and sunspot cycles get progressively farther out of step, on the average one and a half years more for each sunspot cycle. From 1750 to 1792 the lynx peaks and sunspot maxima were within two years of each other; then till 1824 the lynx peaks were close to sunspot minima; until 1850 the previous condition held; until 1877 conditions were reversed; thence to 1906 lynx peaks were within two years of sunspot maxima and recently they have been close to sunspot minima again. This is the situation that necessarily exists with two independent cyclic curves of slightly different period. The sunspot cycle is more variable than the animal cycle, ranging in length from seven to seventeen years, while the animal cycles have varied from eight to twelve years in length. The lags of the lynx peaks after the nearest sunspot peaks average

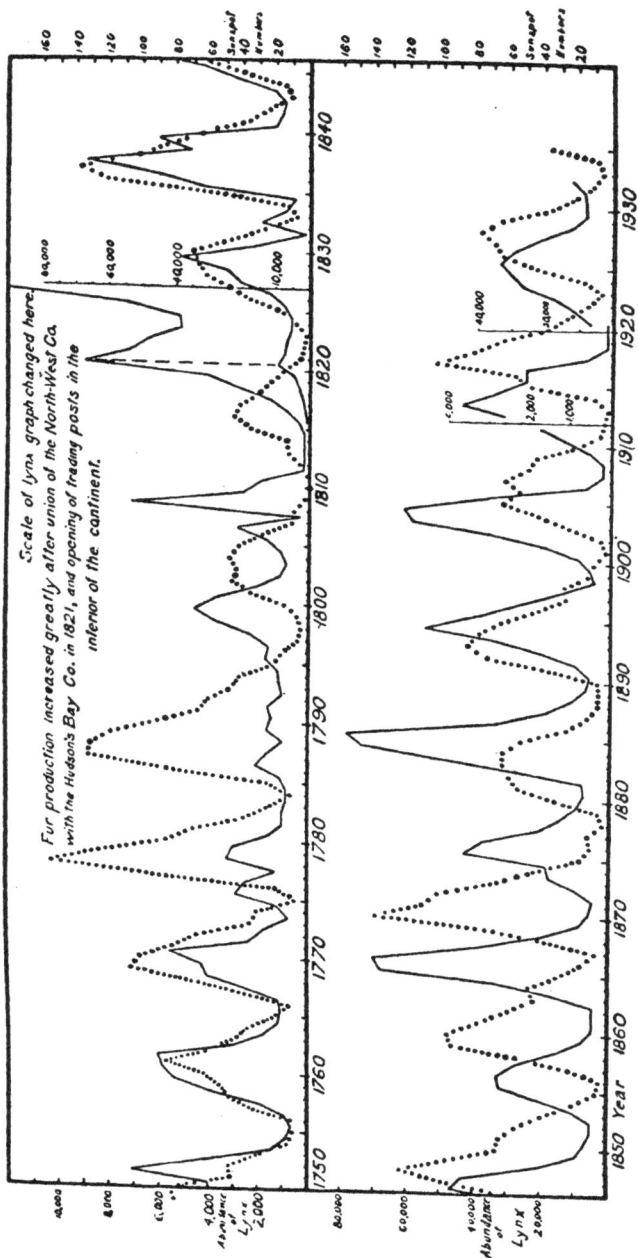

FIGURE 16.—Abundance of lynx (*Lynx canadensis*) compared with sunspot numbers. The lynx abundance in winter (solid line) plotted over end of year in which the winter began, *i.e.*, after the season of its biological production. Sunspot numbers (dotted line) plotted over middle of each year (the marks on base line). Basis of lynx curve: 1750 to 1911—Hudson's Bay Company fur returns up to 1888 from Poland (1892), to 1906 from Seton (1911), and the last five years from Hewitt (1921); 1912 to 1920 from Elton (1933), Hudson's Bay Company records for Mackenzie river district alone; 1920 to 1932 from "Canada Year Books".

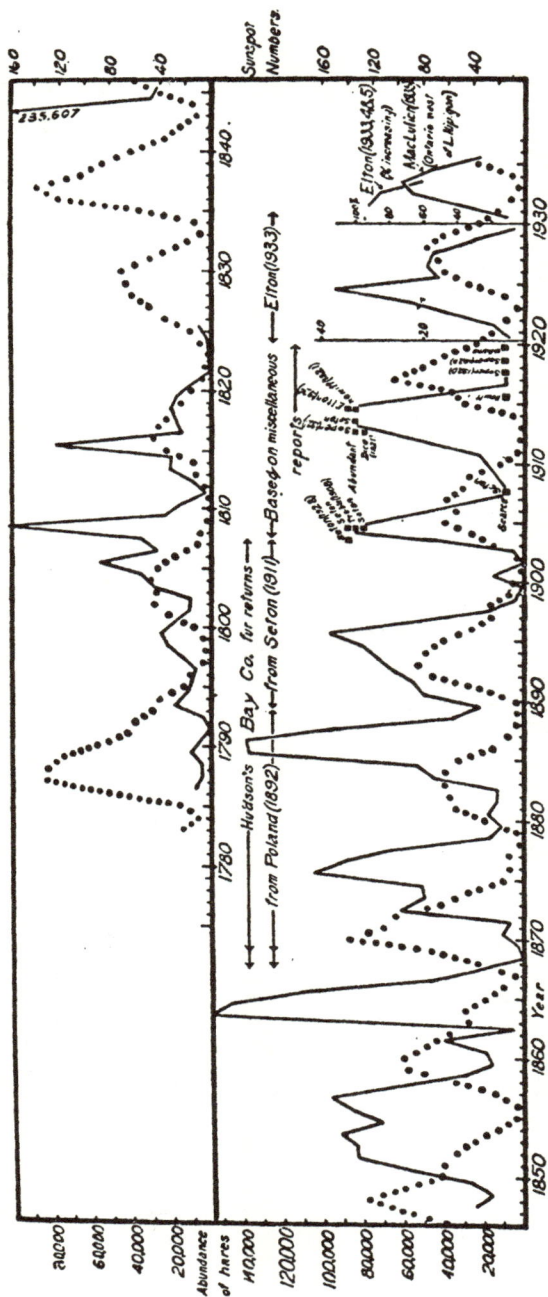

FIGURE 17.—Abundance of varying hare compared with sunspot numbers. The abundance of hares in winter (solid line) plotted over end of year in which the winter began, i.e., after the season of its biological production. Sunspot numbers (dotted line) plotted over middle of each year (the marks on base line).

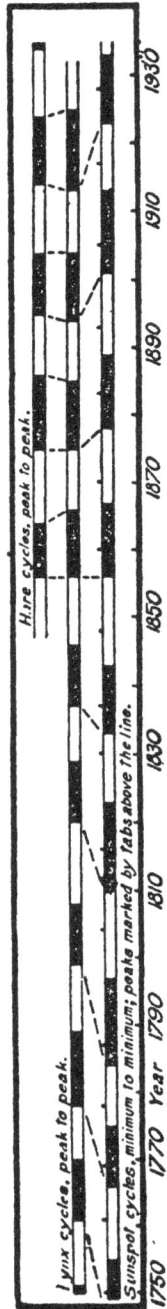

FIGURE 18.—Visual comparison of cycles of abundance of hares, lynxes, and sunspots.   Alternate cycles marked in black.

−1/5 of a year; the lags after the nearest sunspot minima average −1/6 of a year, almost the same amount. Without shifting the curves into phase, as there is no lag shown above, the coefficient of correlation is +0.04 with a standard deviation of ±0.05. In this calculation the lynx abundance was all read off the graph by the one scale of figures running to eighty thousand skins, and the actual sunspot numbers were used. This again shows no correlation between sunspots and lynx numbers.

The fact that lynx abundance is not correlated with sunspot numbers is strong evidence, amounting to proof, that rabbit numbers are not related to sunspots. Between 1856 and 1934 there have been eight rabbit cycles, but only seven sunspot cycles; the average length of the rabbit cycle was 9.7 years, but the sunspot cycle averaged 11.0 years. Therefore the rabbit fluctuation is not related to the sunspot cycle. It has been shown (MacLulich, 1935) that the rabbit peak was reached in 1931 in part of south-eastern Ontario, but not till 1934 in northern Ontario. This situation does not suggest control by the sunspot cycle. The lags of the rabbit peaks after the nearest sunspot minima were (1856-1933): 0, −3, −3, −3, +6, +3, +1, +1, +1, and average −1/3 of a year, but the lags after the nearest sunspot maxima were: −4, +4, +5, +3, +2, −1, −3, −4, +5, and average +7/9 of a year. If the curves were correlated and one lag was small the other would be large, equal to nearly half the length of the cycle. The coefficient of correlation of the Hudson's Bay Company rabbit figures from 1847 to 1903 with sunspot numbers is −0.08 with a probable error of ±0.09; and the coefficient for the whole period from 1847 to 1934 inclusive, is −0.08 with a probable error of ±0.07. This means the rabbit and sunspot curves are not related.

There are recurring periods during which the animal peaks or other phenomena correspond more or less closely to the sunspot maxima, each followed by a period when the animal curve is roughly in phase with the sunspot minima, as was pointed out above for the lynx. Authors have noted this correspondence, especially over periods of fifty years or less. and have assumed a significant correlation. There appears to exist a degree of cyclic correspondence among many animal populations, but it has been shown above that it does not correlate with sunspot numbers in the cases studied, where the history goes back far enough to allow definite conclusions to be drawn. The short, three to four year cycles of lemmings and other mice (Elton, 1924) obviously are not caused by the much longer sunspot cycle.

The conclusion to •be drawn is that the fluctuations in numbers of neither lynx (*lynx canadensis*) nor varying hare (*Lepus americanus*) are correlated with sunspots.

DISCUSSION OF POPULATION PHYSIOLOGY

Considerable experimental work has been done in the study of the growth and regulation of populations, but, at least until recently, this was exceeded in volume by theoretical discussions and speculations. We shall briefly review the facts and concepts of population physiology and make applications to the case of the varying hare.

Growth of Populations, Single Species

Verhulst (1838, 1844) was interested in the problem of the manner of growth of a population and saw that, although growth tended to be geometric, yet the size of a population was limited by the available space and food. He assumed that the limiting effect would make itself felt gradually until, when the population reached the maximum density, the pressure resisting growth would completely counterbalance the tendency to increase. That is, the density attained at any time was that which would result from geometric growth, minus a quantity which depended on the size of the population. For this quantity various functions were discussed and a simple one found to be suitable, namely, the square of the population multiplied by a constant. The relation was expressed in a mathematical formula, and applied to population figures for France, Belgium, the county of Essex in England, and the people of Russia of the Greek Church; with good correspondence between the actual and the calculated figures.

This law of the growth of populations remained in obscurity until it was re-discovered independently by Pearl and

Reed (1920) and it has since been known as the Verhulst-Pearl-Reed Law. A more informative and less empirical form of the equation than Verhulst used is: (using the notation

of Gause, 1934) $\dfrac{dN}{dt} = bN\dfrac{K-N}{K}$; where $b =$ biotic potential or

the possible geometric rate of growth; $N =$ density of population; and $K =$ the maximum population that can exist in a given microcosm with a given level of food resources. In words this equation says:

$$\begin{pmatrix}\text{rate of} \\ \text{growth}\end{pmatrix} = \begin{pmatrix}\text{potential} \\ \text{increase of} \\ \text{population per} \\ \text{unit of time}\end{pmatrix} \times \begin{pmatrix}\text{degree of realization of} \\ \text{the potential increase.} \\ \text{Depends on the number} \\ \text{of still vacant places.}\end{pmatrix}$$

to quote Gause (1934). The deducted quantity of Verhulst

is, in this notation, $\dfrac{b}{K} N^2$. Gause (1934) pointed out that this

logistic curve of Verhulst-Pearl-Reed gave quantitative expression to the ideas of biotic potential and environmental resistance (Chapman, 1928, 1931), and intensity of the struggle for existence.

This equation has been found to fit the growth of the populations of the United States (Pearl and Reed, 1920; Pearl, 1921), the negro population of the United States (Gover, 1929), Sweden, France, Algeria (Pearl, 1925), Mauritius (Anderson, 1929); the fruit fly (*Drosophila*) (Pearl, 1925), the confused flour beetle (*Tribolium confusum*) (Stanley, 1932), several species of protozoa (Gause, 1934), bacterial cultures (Thornton, 1922; Lotka, 1925), several species of yeasts (Slater, 1921; Pearl, 1925; Gause, 1934); and also the growth of some individual organisms such as sunflower seeds (Lotka, 1925).

It seems then that the Verhulst-Pearl logistic curve expresses quantitatively and simply, to a first approximation at least, the struggle for existence which takes place between individuals of a homogeneous group.

Correlation of Density of Population with Other Factors

There is evidence both for and against the idea of high density of population causing a decreased rate of reproduction. Woodruff (1911) found a decreased rate of reproduction with increased density of population of paramecia, Robertson (1921) showed opposite effects with some infusorians, up to certain low densities. Allee (1930) and Johnson (1933) have reviewed the work that was done to check Robertson's findings, and the results were conflicting. Johnson's (1933) results were to the effect that the densities of bacteria, the food, influenced to a considerable degree the rate of reproduction, and perhaps completely controlled it. With the confused flour beetle (Park, 1934) showed that the decreased net rate of reproduction was due to increased cannibalism in part, but also to an inhibitory effect on the rate of egg production of the accumulated products in flour in which beetles had lived. Kendeigh (1934) found an inverse relation between density and reproduction in house wrens (*Troglodytes aedon*).

The general result seems to be that density of population, in the cases studied in greater detail, has not a direct and positive effect on reproduction, but its influence on other factors of the environment will indirectly change the rate of reproduction, and not always in the same direction.

The literature on the harmful effects of crowding animals into a limited environment was reviewed by Allee (1931). That on the beneficial results of crowding has been reviewed by Fowler (1931) in introducing his experiments with *Daphnia longispina*. Groups of them survived longer than single individuals in strong concentrations of toxic electrolytes, but single individuals survived longer than groups in weaker solutions. This was parallelled by the finding that introduction of carbon dioxide into toxic solutions where Daphnia were living increased their survival time in strong concentrations and decreased it in weak concentrations of the electrolytes. Allee (1928, 1929, 1933) demonstrated an unknown protective substance in culture media in which the marine turbellarian flatworm (*Procerodes wheatlandi*) had grown, which increased their survival time in fresh water.

Populations of Several Species

Under this heading we shall first mention some concepts of epidemiology, the struggle for existence of a host species and the disease producing organism. A number of epidemicity indices were summarized by Pearl (1924), of which few are adapted to the meagre data available for epidemics among animals in the wild. A combination of his indices Nos. 4 and 5 gives the following: $\text{Index} = \dfrac{M - M^1}{T}$ where M = mean death rate during epidemic, M^1 = mean death rate over long period, and T = duration of the epidemic. This is a measure of the explosiveness of the epidemic.

The struggle for existence was studied mathematically on paper without experimental confirmation before biologists began to investigate the problem. Ross (1908, 1911) developed equations for the spread of malaria; which expressed the simultaneous dependence of the infection of humans on the amount of infection in mosquitoes, and of infection in mosquitoes on infected humans. An interesting conclusion derived from these equations is that whatever the original number of malaria cases the malaria ratio will settle down to a fixed figure, so long as conditions remain constant; but that if the number of mosquitoes (anophelines) in the district is below a certain value the ultimate malaria rate will tend to zero, that is, the disease will tend to die out in that locality. (See analysis by Lotka 1923, 1925).

A number of combinations of several species living together, some competing for the same places, other cases of competing for the same food, and still other of prey and predator relationships, were discussed and equations formulated by Lotka (1920, 1923a, 1923b, 1925, 1932a, 1932b) and independently by Volterra (1926, 1928, 1931a, 1931b).

The case of two species competing for the same food was experimentally investigated by Gause (1932) using two species of yeasts, and the theoretical equations of Volterra were completely realized. Gause demonstrated that it was possible to predict in what proportion a certain limited amount of food will be distributed between the population of two competing species. With two species of protozoa competing for the same food which was replenished regularly to a definite amount, one species always replaced the other. Further research was needed before the rate of displacement could be predicted, but the qualitative prediction could be made as to which species would displace the other.

The theoretical equations of Lotka and Volterra dealing with the destruction of one species by another have the property of a periodic solution. As the number of predators increases the prey diminish in number, but when the concentration of the latter becomes small the predators, owing to an insufficiency of food, begin to decrease. This produces an opportunity for growth of the prey, which again increase in number.

The discussion was continued by Gause (1934) on the basis of experiments with two species of protozoa, one feeding on the other, *Didinium nasutum* and *Paramecium caudatum*. In this case the "classical" fluctuations did not take place, but instead the prey was all eaten up, resulting then in the death of all the predators. Gause called attention to another type of fluctuation which physiologists and physicists call "relaxation oscillations". With this type of oscillation, a certain condition is built up by a continuous process until a discharge or epidemic (considering the cause of the epidemic as a predator) takes place, which leads to an interruption of contact. These relaxation oscillations described the interaction of the two protozoa.

Gause pointed out that "classical" oscillations are possible around a point called the "singular point" on a plotted map of the curves of relaxation oscillations. The "classical" fluctuations of Lotka and Volterra are really a special case of the relaxation oscillations. However, the zone of possible "classical" oscillations was displaced to such small densities of population with the two protozoa that such fluctuations had no chance of occurring.

To enlarge the zone of possible "classical" fluctuations, Gause (1935) chose *Paramecium* devouring yeast cells, where the intensity of consumption of prey was naturally low. In this case actual fluctuations were obtained.

Conditions more nearly approaching those occurring in nature were tried with the first two protozoa by providing a refuge of sediment in the bottoms of the culture tubes, into which *Paramecium* would go; but *Didinium* would not enter it. The *Didinium* ate the *Paramecium* in the liquid above the sediment, and then all died, whereupon the culture became fully stocked with *Paramecium*. That was what happened in most of the cases, but the refuges were variable, and the number of *Paramecium* that entered into them varied, so that the struggle for existence was controlled by such a multiplicity of causes that it was impossible to predict exactly the course of development of each individual microcosm. "From the language of rational differential equations we are compelled to pass on to the language of probabilities, and there is no doubt that the corresponding mathematical theory of the struggle for existence may be developed in these terms."

By introducing one *Paramecium* and one *Didinium* at equal intervals of time into a microcosm without a refuge, periodic oscillations occurred, not as an inherent property of the predator-prey interaction, but due to the regular interferences. At critical moments, when one cycle of growth succeeded another, the number of individuals being small, multiplicity of causes acquired great significance, and it was impossible to forecast exactly in what way the development would proceed past such a critical point.

Applications to Varying Hares

The present writer now suggests that the relaxation oscillations describe the fluctuations of the varying hare rather than the "classical" equations of Lotka and Volterra.

The graphs of abundance (figs. 10 and 11), especially that for the more limited and homogeneous region around Frontenac county, show a gradual increase to abundance, followed by a sudden sharp decrease. (See also the population figures derived from field observations of hares and their droppings.) It is suggested that, after an epidemic has died out due to immunity and isolation of the survivors, the hares increase from a condition of scarcity approximately according to the logistic curve of Verhulst-Pearl-Reed, tending to reach a nearly constant level of abundance which is maintained for several seasons. The shape of the curves, mentioned above, for the period of increase conforms well to the logistic curve.

The high density of population never endures for long because at that amount of congestion and under the conditions of the animals' sanitation and feeding, the spread of some disease or parasite is facilitated and indeed inevitable. The causative agent of these epizootics may be different at different times and places, depending on many factors, such as the existence locally of some carriers or cases of infection, weather conditions more favourable to one organism than another, and so on. This is a critical time when one organism or another may take the lead in becoming epizootic. That the disease agent is not always the same is shown by the fact that stomach worms killed off hares at Smoky Falls but they were not the cause of decrease at Buckshot lake. Moreover, Green and Shillinger consider tularemia to have killed off hares in Minnesota, but it did not kill them off at Smoky Falls. The varied nature of the reports of symptoms accompanying heavy decreases in hares, and the variation in the time of decreases, indicate that a few wholesale decimations have occurred in winter but the majority in summer. Seton (1909, 1928) wrote: "I feel more and more satisfied that the so-called Rabbit plague is not one disease, but many run riot, through the Rabbits being overcrowded and the whole country rendered unsanitary for their species."

This second part of the hare cycle may be described in terms of the relaxation oscillations. Of course we lack population figures for a whole cycle, and sufficient quantitative knowledge of the biological factors, so that we cannot formulate actual equations. The organism causing the decrease in hare abundance increases in numbers but finally has killed all the vulnerable hares and itself disappears. In nature there are "refuges" to prevent the extinction of either form. For hares these take the form of immunity and isolation after chance survival till hares had become scarce, to mention two possibilities, and for the disease organisms refuges consist in occasional hares which will be affected and so carry the organisms through to the next time of decrease.

The length of the cycle is about the same each time because it depends chiefly on the time the population takes to grow from scarcity to abundance, following the logistic or

sigmoid curve of growth, and secondly on the time required
for the relaxation reaction.   The relaxation oscillation takes
a short time because a slow acting infection would, except by
a rare chance, be outstripped and supplemented by a faster
acting epidemic infection.

In regard to the matter of food supply, some further discussion will be profitable here.   While the hares are increasing
much food exists and they increase to a maximum level of
abundance which would be maintained, barring epidemics,
if the same amount of food that was available when the
hares increased were replenished yearly.   Food is abundantly
produced in summer, but bark and twigs for food in winter
are produced more slowly, so that hares would one winter
be short of food and so decrease to a new level of abundance
for the changed conditions, presumably in accordance with
the steeply descending curve given by Lotka's (1925) formula
13 for a population that initially exceeded its equilibrium
strength.   That this has seldom happened is shown by the
fact that nearly all our recorded dead hares and decreases in
numbers have been found in summer, and in the second place
by the fact that the decrease goes vastly farther than would
be brought about by this factor.   Hares decrease so far that
the food produced is virtually not utilized at all during a
considerable succession of years.   Epidemics have usually
taken hold before the above check operated.

In talking of the population of snowshoe hares and other
animals, certain phrases are often used, such as, "the average
around which the population fluctuates", the "equilibrium
towards which the population tends", and the "balance of
nature".   The last concept may have some value if it carries
only the idea that species rarely become extinct over their
whole range; to that extent there is control.   The words in
themselves imply an essentially stable condition with the
internal forces so proportioned to each other that they do
not tend to change.   In this latter sense the term "balance
of nature" cannot be truly applied to the animal communities
in which the hare is included.   The first two terms above
are commonly synonymous and mean that the population
tends to stay at a value equal to its average size taken over a
period of years, as if attached to that by an elastic cord.

We have seen that the population of hares shows no such effect as it passes the average value, and has no recognizable centre point of balance. It tends to go to the asymptotic maximum value, which instead of the average might then be truly called the "equilibrium towards which the population tends".

In conclusion, a hypothetical reconstruction (table 18) of the population changes of hares during a cycle may be of interest. The following assumptions were used: (1) In the first year after the lowest point reached by the population, females produced only two young per litter; three young the succeeding four years, four young for the year of greatest increase, then three young for one year, and during the time

TABLE 18.—Hypothetical hare cycle. Population figures are totals per square mile.

Years	1	2	3	4	5	6	7	8	9	10
Number of young per litter...	2	3	3	3	3	4	3	2	2	2
Number of litters per season..	2	2	2	2	2	2	2	1	1	1
Breeding stock at beginning of season.................	2	4	12	36	108	324	1,396	4,188	288	18
Half, equal number of females	1	2	6	18	54	162	698	2,094	144	9
Young produced.............	4	12	36	108	324	1,396	4,188	4,188	288	18
Total hares................	6	16	48	144	432	1,720	5,584	8,376	576	36
Normal mortality..........	2	4	12	36	108	324	1,396	4,188	288	18
Epidemic mortality..........	0	0	0	0	0	0	0	3,900	270	16
Remainder.................	4	12	36	108	324	1,396	4,188	288	18	2
Per cent. mortality..........	33	25	25	25	25	19	25	96	97	94

of decrease only two young. (2) Each female had two litters per season, except during years of decrease when only one litter was produced before she died of epidemic disease. (3) The decrease occurred as rapidly as shown by the figures for Buckshot lake given in the results of field work. (4) Hares died at an average age of one year. (Green and Shillinger (1935) found only 5 per cent. survival over one year of age.) This was taken to account for approximately all the normal mortality.

This imaginary cycle had two years of abundance, the seventh and eighth, and during the latter year the epidemic set in, raising the mortality to 71 per cent. higher than normal.

The epidemicity index developed above, namely, the excess mortality above normal divided by the duration of the epidemic, has a value of 24, which signifies an extremely explosive epidemic.

ACKNOWLEDGEMENTS

The writer is pleased to acknowledge the full co-operation and facilities provided by the Department of Biology and the School of Hygiene, University of Toronto, and the Royal Ontario Museum of Zoology. To Professor J. R. Dymond, Director of the Museum, the writer is particularly indebted for his constant interest, help, and advice.

The Ontario Fisheries Research Laboratory generously extended its hospitality and transportation whenever the author worked at Frank's bay, and the assistance of Professor W. J. K. Harkness and the other members of the staff is gratefully acknowledged.

The Ontario Forestry Branch on a number of occasions lent a canoe, and use of a cabin. Facilities and assistance in Algonquin Provincial Park were extended on a number of occasions by the superintendent, Mr. F. A. MacDougall.

Dr. C. H. D. Clarke accompanied the writer during 1932 and 1933, contributing specimens and comradeship.

The visit to Smoky Falls was made possible through the courtesy of the Spruce Falls Power and Paper Company, which provided transportation on its railroad.

The following are to be thanked for identifying sample specimens or carrying out various tests: Dr. M. H. Brown and Miss Anderson, Dr. D. T. Fraser, Dr. C. E. Dolman and his successor Dr. J. S. Kitching, Dr. A. L. MacNabb, Director of the Ontario Public Health Laboratory and Miss Crossley, Dr. R. Matheson of Cornell University and the Taxonomic Division of the Bureau of Biological Survey at Washington.

To all the eight hundred and more persons who contributed information, we extend our sincere thanks.

SUMMARY AND CONCLUSIONS

The object of the study was to investigate the periodic fluctuations in numbers of the varying hare (*Lepus americanus*).

The abundance of hares was determined from four types of material:

(*a*) fur returns,

(*b*) statements in the literature,

(*c*) questionnaires,

(*d*) field work.

In estimating abundance from the fur returns of the Hudson's Bay Company, the year of biological production of the furs has been used, instead of the year of sale, which had been used by other writers. The year of biological production is the second year previous to the year of sale.

A new method of compiling and mapping the reports was used, which does not depend on personal judgment.

The abundance of hares was measured in the field by four methods: trapping, censuses, comparison of numbers of hares observed while travelling known distances, counts of droppings or scatology.

By means of the census obtained at Smoky Falls, the numbers of hares observed and the scatology were approximately calibrated, so that these methods could be used to yield rough estimates of the numbers of hares per square mile of country.

Populations ranged from one per square mile for extreme scarcity to over one thousand for abundance. The highest number was in the north part of Frontenac county in July, 1932, when there were thirty-four hundred per square mile.

The last years of great abundance before the decreases on the Hudson bay watershed were: 1856, 1864, 1875, 1886, 1895, 1914, 1924, and 1934.

The last year of great abundance in Ontario varied from one region to another as follows: 1932, in a small district centring on the north part of Frontenac county; 1933, from Bruce peninsula and southern Algoma to Renfrew county; 1934, in the height of land country from Timiskaming district past lake Nipigon to include Kenora district; 1935, on the northern part of the clay belt; 1934, on the coast of James bay; and 1933 in the southern part of Patricia region. This progression of the phase of the cycle characterized the decrease in the 1920's as well as the recent one.

For Canada as a whole the cycle appeared to reach a peak earliest in the coastal districts of the maritimes and the St. Lawrence valley, the delta of the Mackenzie river, and British Columbia; and latest in the northern parts of the Canadian life zone and the southern parts of the Hudsonian.

Numbers of young per litter and dates of birth have been recorded in southern Ontario; some young were born at the end of April, but in the more northern part of the province the first litter came at the end of May. There was a second major group of litters in late June, followed by a lesser one in July. A few hares were born as late as September. The average number of young per litter was 3.4, the smallest number one, and the highest number seven.

Three young were born and raised in captivity, and their rates of growth measured.

The body temperature of two hares was found to average 102.3° F., with a range of variation of plus or minus a half degree.

Twelve hares on an area of eight and a half acres at Smoky Falls ate the bark off a hundred and sixty-eight poplars and a few other trees and shrubs during the winter of 1934-5.

Hares were found entering and using holes in the snow made by themselves, and burrows in the ground which may or may not have been made by the hares. This habit was observed to be usual with hares in northern Ontario. Hares were found in surface "forms" on a few occasions only.

Several observations on the range of movement of hares were obtained, the greatest being two hundred and fifty yards.

The commonest or "typical" habitat in Ontario is coniferous swamps, and in western Canada alder-willow swamps and thickets.

Population figures for the northern part of Frontenac county showed that in times of abundance the hares occupied practically all types of habitat whereas in times of scarcity they were confined to their favourite habitat.

A number of the intestinal bacteria found in "normal" varying hares have been identified, and a rough estimate of their frequency of occurrence obtained.

Salmonella suipestifer (Kruse) Lignières, which causes disease in swine, was isolated from two varying hares, the first record of its occurrence in an animal in the wild state.

A fatal diseased condition due to *Pneumococcus Type* XIX was described, the strain having been obtained, apparently at least, from rabbit ticks (*Haemaphysalis leporis-palustris*).

Infection by *Staphylococcus aureus* has been shown to be common; occasionally with fatal results. Hares have practically no *Staphylococcus* antitoxin.

By serum tests it was shown that tularemia occurred among varying hares in widely separated parts of Ontario, but not in any large percentage of the population.

There was no evidence that any of these bacterial diseases was causing epidemics.

Ten species of helminths were found in the hares examined.

The blood-sucking stomach worms, *Obeliscoides cuniculi*, caused an epidemic at Smoky Falls on the Mattagami river in the summer of 1935. Seven recently caught and well-cared-for hares died, six of them due definitely to excessive infection with this parasite.

Records of occurrence were secured for three mites, two ticks, one flea, and three other insects.

The decrease in abundance of hares was due to wholesale dying-off.

Questionnaire replies suggest that the decrease in the 1920's was due to a more severe and noticeable epidemic than was the recent decrease.

The epidemic is not always the same disease at every time and place.

Data on the rate of reproduction bear out the statements of other naturalists to the effect that the rate is decreased at the time of decrease in abundance of hares; but it could not account for the cycle.

The abundance of lynx was shown to be definitely correlated with that of varying hares, as a conclusion from a more extended and accurate analysis than had been made before.

No correlation was found between rainfall and decrease in hare numbers.

It was demonstrated that the fluctuations in numbers of neither lynx nor varying hares are correlated with sunspots.

It was concluded that the relaxation oscillations, applied by Gause (1934) to animal populations, describe the fluctuations of the hare, rather than the "classical" equations of Volterra.

LITERATURE CITED

Alicata, J. E. 1932. Life history of the rabbit stomach worm, *Obeliscoides cuniculi.* Jour. Agric. Res., **44**: 401-420.

Allee, W. C. 1928. Studies in animal aggregations: Mass protection for Procerodes, a marine turbellarian. Jour. Exp. Zool., **50**: 61-84.

Allee, W. C. 1929. Studies in animal aggregations: Mass protection from hypotonic sea-water for Procerodes, a marine turbellarian. Jour. Exp. Zool., **54**: 349-379.

Allee, W. C. 1931. Animal aggregations. University of Chicago Press.

Allee, W. C. 1933. Studies in animal aggregations: Further analysis of the protective value of biologically conditioned fresh water for the marine turbellarian, *Procerodes.* Physiol. Zool., **6**: 1-21.

Anderson, D. D. 1929. The point of population saturation: Its transgression in Mauritius. Human Biology, **1**: 528-543.

Anderson, R. M. 1932. Methods of collecting and preserving vertebrate animals. Nat. Mus. Canada, Bull. 69.

Anonymous. 1935. The seasonal food habits of the snowshoe hare. Forest Research Digest (Lake States Forest Experiment Station), May, p. 7, mimeographed.

Anonymous. 1936. Woody food preferences of the snowshoe rabbit in the lake states. Technical Notes (Lake States Forest Experiment Station), multilithed.

Anthony, H. E. 1934. Sun-spots and squirrel emigrations. Literary Digest, Jan. 20, **11**: 30-31.

Audubon, J. J. and Bachman, J. 1851 to 1856. Quadrupeds of North America. New York.

Bailey, V. 1921. Capturing small mammals for study. Jour. Mamm., **2**: 63-68.

Bailey, V. 1932. Trapping animals alive. Jour. Mamm., **13**: 337-342.

Bailey, V. 1933. Live traps for game farms. The Game Breeder, **37**: 2-3.

Bailey, V. 1934. Directions for setting foothold trap for capturing animals uninjured. Bur. Biol. Surv., Bi-1332, June.

Bergey, D. H. 1934. Fourth Edition, Bergey's manual of determinative bacteriology. Published at the direction of the Soc. Amer. Bacteriologists, Baltimore.

Boughton, R. V. 1932. The influence of helminth parasitism on the abundance of the snowshoe rabbit in western Canada. Can. Jour. Res., **7**: 524-547.

Brückner, Edouard. 1890. Klimaschwankungen seit 1700 nebst Bemerkungen über die Klimaschwankungen der Diluvialzeit. Wien.

Buchanan, A. 1920. Wild life in Canada. Toronto.

Burt, W. H. 1927. A simple live trap for small mammals. Jour. Mamm., **8**: 302-304.

Cahn, A. R., Thomas, L. J., and Wallace, G. I. 1932. A new disease of moose, III. A new bacterium. Science, **78**: 385-386.

Chapman, R. N. 1928. The quantitative analysis of environmental factors. Ecol., **9**: 111.

Chapman, R. N. 1931. Animal Ecology. New York.

Chapman, R. N. 1933. Annual Address, The causes of fluctuations of populations of insects. Proc. Hawaiian Ent. Soc., 8: 279-292.

Chapman, R. N. and Baird, Lilian. 1934. The biotic constants of *Tribolium confusum* Duval. Jour. Exp. Zool., 68: 293-304.

Christy, Miller. 1892. Extermination of the rabbit in Australia. Zoologist, p. 383.

Clarke, C. H. D. 1936. Fluctuations in numbers of the ruffed grouse, *Bonasa umbellus* (Linné), with special reference to Ontario. Univ. Toronto Studies, Biol. 41.

Clemow, F. G. 1900. Jour. Trop. Med., 2: 169.

Clopper, C. J. and Pearson, E. S. 1934. The use of confidence or fiducial limits illustrated in the case of the binomial. Biometrika, 26: 404-413.

Collett, R. 1895. Christiania Vidensk-Selsk. Forhandl. No. 3, 52.

Collett, R. 1911-12. Norges Pattedyr, p. 64.

Coventry, A. F. 1931. Amphibia, reptilia and mammalia of the Temagami district, Ontario. Can. Field-Nat., 45: 109-113.

Cowdry, E. V. and Ham, A. W. 1932. Studies on East Coast fever. (1) The life cycle of the parasite in ticks. Parasitology, 24: 1-49.

Criddle, Norman. 1932. The correlation of sunspot periodicity with grasshopper fluctuation in Manitoba. Can. Field-Nat., 46: 195-199.

Cross, E. C. and Dymond, J. R. 1929. The mammals of Ontario. Roy. Ont. Mus. Zool., Handbook No. 1.

Curtice, C. 1892. Parasites, being a list of those infesting the domesticated animals and man in the United States. Jour. Comp. Med. and Vet. Arch., 13: 223-236.

Danysz, J. 1893. Jour. Agric. Pratique, 2: 920.

Danysz, J. 1900. Un microbe pathogène pour les rats (*Mus decumanus* et *mus ratus*) et son application à la destruction de ces animaux. Ann. Inst. Past., 14: 193-201.

Darwin, Charles. 1859. The origin of species by means of natural selection.

Day, A. M. and Shillinger, J. E. 1935. Predators and rodents are factors in the spread of disease. U.S. Dept. Agric. Yearbook, pp. 284-286.

Dearborn, Ned. 1932. Foods of some predatory fur-bearing animals in Michigan. School Forestry and Conservation, Univ. Mich., Bull. 1.

DeLury, Ralph E. 1923. Arrival of birds in relation to sunspots. Auk, 40: 414-419.

DeLury, Ralph E. 1930. Sunspots in relation to fluctuations in grasshoppers and grouse at Aweme, Manitoba. Can. Field-Nat., 44: 120.

DeLury, Ralph E. 1931. Sunspots and forest life. Illustrated Can. Forest and Outdoors, 27: 314.

Dice, L. R. 1921. Notes on the mammals of interior Alaska. Jour. Mamm., 2: 20-28.

Dice, L. R. 1925. A live-trap for small mammals. Jour. Mamm., 6: 202.

Dice, L. R. 1931. Methods of indicating the abundance of mammals. Jour. Mamm., 12: 376-381.

Dolman, C. E. 1933. Treatment of localized staphylococcic infections with staphylococcus toxoid. Jour. Amer. Med. Ass., 100: 1007.

Dolman, C. E. 1934. Staphylococcus antitoxic serum in the treatment of acute staphylococcal infections and toxemias. Can. Med. Ass. Jour., **30**: 601, and **31**: 1 and **31**: 130.

Dominion Bureau of Statistics. 1922 to 1935. Canada Year Books, for 1920 to 1934-35. Ottawa.

Dymond, J. R. 1928. The mammals of the lake Nipigon region, Ontario. Trans. Roy. Can. Inst., **16**: 233-291. Re-issued as Roy. Ontario Mus. Zool., Contrib. 1.

Eidgenössische Sternwarte in Zürich. 1934. Observed relative sunspot numbers, 1749-1933. Terrestrial Magnetism and Atmospheric Electricity, **39**: 231-236. (1934 and 1935 figures compiled from later numbers of this journal.)

Eidmann, H. 1931. Zur Kenntniss der Periodizität der Insektenepidemien. Zeitschr. f. angen. Entom., **18**: 537-567.

Elton, C. S. 1924. Periodic fluctuations in the numbers of animals: Their causes and effects. Brit. Jour. Exp. Biol., **2**: 119-163.

Elton, C. S. 1925. Plague and the regulation of numbers in wild animals. Jour. Hyg., **24**: 138-163.

Elton, C. S. 1927. Animal Ecology. London.

Elton, C. S. 1931a. The study of epidemic diseases among wild animals. Jour. Hyg., **31**: 435-456.

Elton, C. S. 1931b. Epidemics among sledge dogs in the Canadian arctic and their relation to disease in the arctic fox. Can. Jour. Res., **5**: 673-692.

Elton, C. S. et al. 1931c. The health and parasites of a wild mouse population. Proc. Zool. Soc., Lond., pp. 657-721.

Elton, C. S. 1933a. Matamek Conference on Biological Cycles—Abstract of papers and discussions. Matamek Factory, Canadian Labrador.

Elton, C. S. 1933b. The Canadian snowshoe rabbit inquiry, 1931-32. Can. Field-Nat., **47**: 63-69, 84-86.

Elton, C. S. 1934. The Canadian showshoe rabbit enquiry, 1932-33. Can. Field-Nat., **48**: 73-78.

Elton, C. S. 1935. The Canadian snowshoe rabbit enquiry, 1933-34. Can. Field-Nat., **49**: 79-85.

Elton, C. S. 1936. The Canadian snowshoe rabbit inquiry, 1934-35. Can. Field-Nat., **50**: 71-81.

English Plague Commission, Report, 1906.

Erxleben. 1777. Syst. Regni Anim., vol. 1.

Ewing, H. E. 1929. A manual of external parasites. Baltimore.

Fisher, R. A. 1930. Statistical methods for research workers. London.

Fowler, J. R. 1931. The relation of numbers of animals to survival in toxic concentrations of electrolytes. Physiol. Zool., **4**: 214-245.

Francis, E. 1919. Deer-fly fever: A disease of man of hitherto unknown etiology. Public Health Report, **34**: 2061-2062.

Francis, E. 1925. Tularemia. Jour. Amer. Med. Ass., **84**: 1243.

Francis, E. 1928. A summary of present knowledge of tularemia. Medicine, **7**: 411-432.

Francis, E. and Mayne, B. 1921. Experimental transmission of tularemia by flies of the species *Chrysops discalis*. Public Health Report, **36**: 1738.

Francis, E. and Moore. 1926. Identity of Ohara's disease and tularemia. Jour. Am. Med. Ass., 86: 1329-1332.

Galli-Vallerio, B. 1900. Les puces des rats et des souris jouent-elles un rôle important dans la transmission de la peste bubonique à l'homme. Centralb. Bakt. Parasit. Infekt. Originale, 27: 1.

Galli-Vallerio, B. 1903. Les nouvelles recherches sur l'action des puces des rats et des souris dans la transmission de la peste bubonique. Centralb. Bakt. Parasit. Infekt. Referate, 33: 753.

Gause, G. F. 1932. Experimental studies on the struggle for existence. I. Mixed population of two species of yeast. Brit. Jour. Exp. Biol., 9: 389.

Gause, G. F. 1934. The struggle for existence. Baltimore.

Gause, G. F. 1935. Experimental demonstration of Volterra's periodic oscillations in the numbers of animals. Brit. Jour. Exp. Biol., 12: 44-48.

Gauthier, J. C. and Raybaud, A. 1902. Sur le rôle des parasites du rat dans la transmission de la peste. Comptes Rend. Soc. Biol., 54: 1497.

Gauthier, J. C. and Raybaud, A. 1903. Rev. d'Hyg., 25: 426.

Gover, Mary. 1929. Increase of the negro population in the United States. Human Biology, 1: 263-273.

Green, H. U. 1932. Mammals of the Riding Mountain National Park, Manitoba. Can. Field-Nat., 46: 149-152.

Green, R. G. 1925. Distemper in the silver fox. Proc. Soc. Exp. Biol. and Med., 22: 546-548.

Green, R. G., Ziegler, N. R., Green, B. B., and Dewey, E. T. 1930. Epizootic fox encephalitis. 1. General description. Amer. Jour. Hyg., 12: 109-129.

Green, R. G. 1931a. Progress in the study of tularemia. Trans. Eighteenth Amer. Game Conference, p. 252.

Green, R. G. 1931b. Epizootic encephalitis of foxes. II. General consideration of fur-range epizootics. Amer. Jour. Hyg., 13: 201-223.

Green, R. G. 1931c. The occurrence of Bacterium tularense in the eastern wood tick, Dermacentor variabilis. Amer. Jour. Hyg., 14: 600-613.

Green, R. G. and Shillinger, J. E. 1932. Relation of disease to wildlife cycles. Trans. Nineteenth Amer. Game Conference, pp. 432-436.

Green, R. G. and Shillinger, J. E. 1934. Progress report of wildlife disease studies for 1933. Trans. Twentieth Amer. Game Conference, pp. 288-297.

Green, R. G. and Shillinger, J. E. 1935. Progress report of wildlife disease studies for 1934. Trans. Twenty-first Amer. Game Conference, pp. 397-401.

Greenwood, M. and Topley, W. W. C. 1925. A further contribution to the experimental study of epidemiology. Jour. Hyg., 24: 45.

Grinnell, Joseph. 1914. An account of the mammals and birds of the lower Colorado valley with especial reference to the distributional problems presented. Univ. Calif. Publ. Zool., 12: 51-294; p. 92.

Hadwen, S. 1913. On "tick paralysis in sheep and man following bites of Dermacentor venustus; with notes on the biology of the tick". Parasitology, 6: 283-287 and 6: 298.

Hall, M. C. 1916. Nematode parasites of mammals of the orders Rodentia, Lagomorpha and Hyracoidea. Proc. U.S. Nat. Mus., 50: 1-258.

Hall, M. C. 1919. The adult taenoid cestodes of dogs and cats, and of related carnivores in North America. Proc. U. S. Nat. Mus., 55: 1-94.

Hanns, J. V. 1932. Handbuch der Klimatologie. 4th ed., Stuttgart, pp. 408-411.

Hardy, Manly. 1910. A fall fur hunt in Maine. Forest and Stream, May 14, p. 769 and May 28, p. 850.

Harkin, J. B. The fluctuation in the abundance of rabbits. Can. Field-Nat., 41: 113.

Hatt, R. T. 1925. A new livetrap for field use. Jour. Mamm., 6: 178-181.

Hewitt, C. G. 1915. A contribution to a knowledge of Canadian ticks. Trans. Roy. Soc. Canada, 9: 225-239.

Hewitt, C. G. 1921. The conservation of the wild life of Canada. New York

Hind, H. Y. 1860. Narrative of the Canadian Red river exploration expedition, vol. 1.

Horne, H. 1912. Eine Lemmingpest und eine Meerschweinchenepizootie. Ein Beitrag zur Beleuchtung der Ursachen der Lemmingsterbe in den sogenannten Lemmingjahren. Centralb. Bakt. Parasit. Infekt. Originale, 66: 169.

Hudson, H. D. L. 1930. Preliminary report on a case of tularemia. Can. Med. Ass. Jour., 22: 678.

Huntington, E. 1932. Matamek Conference on Biological Cycles. Report. Matamek Factory, Canadian Labrador.

Huntsman, A. G. 1931. The maritime salmon of Canada. Biol. Board of Canada, Bull. 21.

Huxley, Julian. 1927. Fluctuations of mammals and the sun-spot cycle. Harpers Magazine, Dec., pp. 42-50.

Johnson, W. H. 1933. Effects of population density on the rate of reproduction in Oxytrichia. Physiol. Zool., 6: 22-54.

Keane, C. 1927. The outbreak of foot and mouth disease among deer in the Stanislaus National Forest. Calif. State Dept. Agric. Monthly Bull., 16: 213-225.

Kendeigh, S. C. 1934. The role of environment in the life of birds. Ecol. Monographs, 4: 299-417.

Kitasato, S. 1894. The bacillus of bubonic plague. Lancet, 2: 428.

Lake, and Francis, E. 1922. Public Health Report, 37: 392.

Law, R. G. 1933. Tapeworm in rabbits. Can. Field-Nat., 47: 142.

Leopold, Aldo. 1931. Report on a game survey of the north central states. Madison, Wisconsin.

Leopold, Aldo and Ball, John N. 1931. British and American grouse cycles. Can. Field-Nat., 45: 162-167.

Leopold, Aldo. 1933. Game management. London and New York.

Ljungman, Axel. 1880. Bidrag till lösningen af frågan om de stora sillfiskenas sekulära periodicitet. Copenhagen. Translated by Herman Jacobson in report of the commissioner for 1879, U.S. Comm. of Fish and Fisheries, part 7, in 1882, pp. 497-504.

Loeffler, F. 1886. Arbeiten aus dem Kaiserlichen Gesundheitsamte, 1: 141.

Loeffler, F. 1892. Ueber Epidemieen unter den im hygienischen Institute zu Greifswald gehaltenen Mäusen und über die Bekampfung der Feldmausplage. Centralb. Bakt. Parasit. Originale, 11: 129.

Lotka, A. J. 1920. Analytical note on certain rhythmical relations in organic systems. Proc. Nat. Acad. Sc., **6**: 410.

Lotka, A. J. 1923a. Contribution to quantitative parasitology. Jour. Wash. Acad. Sc., **13**: 152.

Lotka, A. J. 1923b. Contribution to the analysis of malaria epidemiology. Amer. Jour. Hyg., **3**, January supplement.

Lotka, A. J. 1925. Elements of physical biology. Baltimore.

Lotka, A. J. 1932a. Contribution to the mathematical theory of capture. I. Conditions for capture. Proc. Nat. Acad. Sc., **18**: 172.

Lotka, A. J. 1932b. The growth of mixed populations; Two species competing for a common food supply. Jour. Wash. Acad. Sc., **22**: 461.

MacConkey, A. T. 1900. Note on a new medium for the growth and differentiation of the Bacillus coli communis and the Bacillus typhi abdominalis. Lancet, part 2, July 7, p. 20.

McCoy, G. W. 1910. Plague among ground squirrels in America. Jour. Hyg., **10**: 589-601.

McCoy, G. W. 1911. A plague-like disease of rodents. U.S. Public Health Service Bull. 43.

McCoy, G. W. and Chapin, C. W. 1912a. *Bacterium tularense*, the cause of a plague-like disease of rodents. U.S. Public Health Service Bull. 53.

McCoy, G. W. and Chapin, C. W. 1912b. Further observations on a plague-like disease of rodents, with a preliminary note on the causative agent, *Bacterium tularense*. Jour. Infect. Dis., **10**: 61-72.

MacFarlane, R. 1890. Canadian Record of Science, **4**: 32.

MacFarlané, R. 1905. Notes on mammals collected and observed in the northern Mackenzie river district, Northwest Territories of Canada, with remarks on explorers and explorations of the far north. Proc. U.S. Nat. Mus., **28**: 673-764. Reprinted in: Mair, Chas., and MacFarlane, R., 1908, Through the Mackenzie basin: A narrative of the Athabaska and Peace River Treaty Expedition of 1899, pp. 153-283. Toronto.

MacLulich, D. A. 1935. Fluctuations in the number of snowshoe rabbits. Forestry Chronicle, **11**: 283-286.

MacLulich, D. A. 1936a. Fluctuations in numbers of snowshoe hares. Science, n.s., **83**: 162.

MacLulich, D. A. 1936b. Mammals of the Wanapitei Provincial Forest, Sudbury district, Ontario. Can. Field-Nat., **50**: 56-58.

MacLulich, D. A. 1936c. Sunspots and the abundance of animals. Jour. Roy. Astronomical Soc. Canada, **30**: 233-246.

MacNabb. A. L. 1930. Tularemia. The first case reported in Canada. Can. Pub. Health Jour., **21**: 91-92.

Macoun, John. 1882. Manitoba and the great North-west, pp. 324-353. Mammals of the north-west. Guelph.

Martelli, G. 1919. Contributo alla concoscenza della vita e dei costumi delle Arvicole in Puglia (Portici).

Meggitt, F. J. 1924. The cestodes of mammals. London.

Miller, Gerrit S., Jr. 1897. Notes on the mammals of Ontario. Proc. Boston Soc. Nat. Hist., **28**: 1-44.

Mitchell, J. A. and others. 1930. Epizootic among veld rodents in De Aar and neighbouring districts of the northern Cape province. Jour. Hyg., **29**: 394-414.

Nicholson, A. J. 1933. The balance of animal populations. Jour. Animal Ecol., **2**: 132-178.

Nikanarov, S. M. 1928. Tularemia in North America and tularemia-like disease in U.S.S.R. Rev. de Microbiol., d'Epidemiol. et de Parasitol, **7**: 289-293.

Nuttall, G. H. F. 1898. Zur Aufklärung der Rolle, welche stechende Insekten bei der Verbreitung von Infektionskrankheiten spielen. Centralb. Bakt. Parasit. Infekt. Originale, **23**: 625.

Nuttall, G. H. F. and Warburton, G. 1915. Ticks: A monograph of the Ixodiodea. Part III, The genus Haemophysalis. Cambridge.

Ogata. 1897. Ueber die Pestepidemie in Formosa. Centralb. Bakt. Parasit. Infekt. Originale, **21**: 769.

Packard, A. S. 1869. List of hymenopterous and lepidopterus insects collected by the Smithsonian expedition to South America, under Prof. James Orton; Appendix to report on Articulates, Ann. Rept. Peabody Academy of Science, pp. 1-14.

Park, Thomas. 1932. Studies in population physiology; the relation of numbers to initial population growth in the flower beetle *Tribolium confusum* Duval. Ecol., **13**: 172-181.

Park, Thomas. 1933. Studies in population physiology. II. Factors regulating initial growth of *Tribolium confusum* populations. Jour. Exp. Zool., **65**: 17-42.

Park, Thomas. 1934. Studies in population physiology. III. The effect of conditioned flour upon the productivity and population decline of *Tribolium confusum*. Jour. Exp. Zool., **68**: 167-182.

Parker, R. R., Spencer, R. R., and Francis, E. 1924. Tularemia infection in ticks of the species *Dermacentor andersoni* Stiles, in the Bitterroot valley, Montana. Public Health Report, **39**: 1057-1073.

Parker, R. R. and Spencer, R. R. 1925. Tularemia and its occurrence in Montana. Sixth biennial report, Montana State Board of Entomology, pp. 30-41.

Parker, R. R. and Spencer, R. R. 1926. Hereditary infection of tularemia by the wood tick, *Dermacentor andersoni* Stiles. Public Health Report, **41**: 1403.

Parker, R. R., Brooks, C. S., and Marsh, Hadleigh. 1929. The occurrence of *Bacterium tularense* in the wood tick, *Dermacentor occidentalis*, in California. Public Health Report, **44**: 1299-1300.

Parker, R. R. 1929. Rocky Mountain Spotted Fever. Ann. Rept. Montana State Board Entom., **7**: 39-62.

Parker, R. R., Hearle, E., and Bruce, E. A. 1931. The occurrence of tularemia in British Columbia. Public Health Report, **46**: 45-46.

Pearl, R. and Reed, L. J. 1920. On the rate of growth of the population of the United States since 1790 and its mathematical representation. Proc. Nat. Acad. Sc., **6**: 275-288.

Pearl, R. 1921. The biology of death— VII, Natural death, public health, and the population problem. Scientific Monthly, pp. 193-213.

Pearl, R. 1924. Studies in human biology. Baltimore.

Pearl, R. 1925. The biology of population growth. New York.

Pearse, R. A. 1911. Insect bites. Northwest Med., n.s., 3: 81-82.

Pirie, H. 1927. Pub. South African Inst. Med. Res., 3: 138 and 163.

Poland. . 1892. Fur-bearing animals in nature and in commerce. London.

Pound, C. J. 1897. The destruction of rabbits by means of the microbes of chicken cholera. Agric. Gaz. N.S. Wales, 8: 538-573.

Preble, E. A. 1902. A biological investigation of the Hudson bay region. North Amer. Fauna, no. 22, U.S. Dept. Agric.

Preble, E. A. 1908. A biological investigation of the Athabaska-Mackenzie region. North Amer. Fauna no. 27. U.S. Dept. Agric.

Richardson, John. 1829. Fauna Boreali-Americana; or the zoology of the northern parts of British America, part 1st, the quadrupeds. London.

Ricker, Wm. E. 1937a. Statistical treatment of sampling processes useful in the enumeration of plankton organisms. Archiv. für Hydrobiologie, 31: 68-84.

Ricker, Wm. E. 1937b. The concept of confidence or fiducial limits applied to the Poisson frequency distribution. Jour. Amer. Statistical Assoc., 32: 349-356.

Ritchie, J. 1926. Trans. and Proc. Perthshire Soc. Nat. Sc., 8: 156.

Robertson, T. B. 1921a. Experimental studies on cellular multiplication. I. The multiplication of isolated Infusoria. Biochem. Jour., 25: 595-611.

Robertson, T. B. 1921b. Experimental studies on cellular multiplication. II. The influence of mutual contiguity upon the reproductive rate of Infurosia. Biochem. Jour., 25: 612-619.

Rosenbach. 1884. Mikroorganismen bei den Wundinfectionskrankheiten des Menschen. Wiesbaden.

Ross, R. 1908. Report on the prevention of malaria in Mauritius.

Ross, R. 1911. The prevention of malaria. London.

Roubakine, A. 1930. Monthly Epid. Rept. League of Nations, R.E. no. 134.

Ruhl, H. A. 1932. Methods of appraising the abundance of game species over large areas. Trans. Nineteenth Amer. Game Conference, pp. 442-450.

Saunders, W. E. 1932. Notes on the mammals of Ontario. Trans. Roy. Can. Inst., 18: 271-309.

Scheffer, T. H. 1934. Hints on live trapping. Jour. Mamm., 15: 197.

Seton, E. T. 1909. Life histories of northern animals; an account of the mammals of Manitoba, 2 vols. London.

Seton, E. T. 1911. The arctic prairies. Toronto.

Seton, E. T. 1925-28. Lives of game animals. 4 vols. 4th, 1928; New York. Reprinted in 8 parts, 1929.

Sharpe, J. F. and Brodie, J. A. 1931. The forest resources of Ontario. Ontario Forestry Branch, Toronto.

Shillinger, J. E. 1933. The significance of wildlife diseases. Proc. 5th. Pacific Science Congress, 4: 2977-2980.

Shillinger, J. E. 1935. Infectious diseases as a cause of loss in wildlife. Louisiana Conservation Review, Oct., pp. 38-40 and 48.

Silver, J. 1929. How to make a cat trap. Leaflet no. 50, Bur. Biol. Surv., Washington (revised 1930).

Simond, P. L. 1898. La propagation de la peste. Ann. Inst. Past., 12: 625-687.

Simroth, H. 1908. Der Einfluss der letzten Sonnenfleckenperiod auf die Tierwelt. Kosmos, 9.

Slator, A. 1921. Yeast crops and the factors which determine them. Trans. Chem. Soc., 119: 115-131.

Snyder, L. L. and Baillie, J. L. 1923. Notes on the birds and mammals of Brent and vicinity, Algonquin Park, Ontario, July and August, 1922. Can. Field-Nat., 37: 89-94.

Snyder, L. L. 1928. The mammals of the lake Abitibi region. Univ. Toronto Studies, Biol. 32; re-issued in Contrib. no. 2, Roy. Ont. Mus. Zool.

Snyder, L. L. 1930. The mammals of King township. Trans. Roy. Can. Inst., 17: 173-181; re-issued in Contrib. no. 3, Roy. Ont. Mus. Zool.

Soper, J. D. 1920. Notes on the mammals of Ridout, district of Sudbury, Ontario. Can. Field-Nat., 34: 61-69.

Soper, J. D. 1921. Notes on the snowshoe rabbit. Jour. Mamm., 2: 101-108.

Soper, J. D. 1922. A biological reconnaissance of portions of Nipissing and Timiskaming districts, northern Ontario. Can. Field-Nat., 36: 175-176, and 37: 11-13.

Soper, J. D. 1923. Mammals of Wellington and Waterloo counties, Ontario. Jour. Mamm., 4: 244-252.

Stanley, J. 1932. A mathematical theory of the growth of populations of the flour beetle *Tribolium confusum* Duv. Can. Jour. Res., 6: 632-671.

Stiles, C. W. 1897. A revision of the adult tapeworms of hares and rabbits. Proc. U.S. Nat. Mus., 19: 145-235.

Swinton, A. H. 1878. Data obtained from solar physics and earthquake commotions applied to elucidate locust multiplication and migration. (Not seen. Possibly in Rept. U.S. Ent. Comm. for the year 1877 relating to the Rocky Mountain locust, *etc.*, Washington.)

Taylor, W. P. 1930. Methods of determining rodent pressure on the range. Ecol., 11: 523-542.

Thiøtta, T. 1930. Nordisk Med. Tidskr., p. 177.

Thiøtta, T. 1931. Tularemia in Norway. Jour. Infect. Dis., 49, Aug.

Thomas, L. J. and Cahn, A. R. 1932. A new disease of moose, I. Preliminary report. Jour. Parasit., 18: 219-231.

Thomson, Andrew. 1936. Sunspots and weather forecasting in Canada. Jour. Roy. Astronomical Soc. Canada, 30: 215-232.

Thornton, H. G. 1922. Annals Appl. Biol., p. 265.

Topley, W. W. C. 1919. The spread of bacterial infection. Lancet, 197: 1-5.

Topley, W. W. C. and Wilson, G. S. 1931. The principles of bacteriology and immunity, 2 vols. London.

Tuttle, L. and Satterly, J. 1925. The theory of measurements. London.

Vail, D. T. 1914. *Bacillus tularense* infection of the eye. Ophth. Rec., 23: 487.

Verhulst, P. F. 1838. Notice sur loi que la population suit dans son accroissement. Corr. math. et phys., pub. par. A. Quetelet, 10: 113-121.

Verhulst, P. F. 1844. Mem. Acad. Roy. Bruxelles, 18: 1, and 1846, 20: 1.

Volterra, Vito. 1926. Variazioni e fluttuazioni del numero d'individui in specie animali conviventi. Mem. R. Accad. Naz. dei Lincei, ser. VI, vol. 2. Translation of large part of it into English in Chapman's Animal Ecology (1931), from a reprint, 1928, in Jour. du Conseil international pour l'exploration de la mer, III, vol. 1.

Volterra, Vito. 1931. Leçons sur le théorie mathématique de la lutte pour la vie. Cahiers scientif. no. 7, Gauthier-Villars et cie. Paris.

Vorhies, C. T. and Taylor, W. P. 1933. The life histories and ecology of Jack rabbits *Lepus alleni* and *Lepus californicus* ssp., in relation to grazing in Arizona. Tech. Bull. 49, Univ. Arizona, College Agric., Agric. Exp. Station, Tucson, Arizona.

Wallace, G. I., Thomas, L. J., and Cahn, A. R. 1932. A new disease of moose, II. Proc. Soc. Exp. Biol. and Med., 29: 1098-1100.

Wallace, G. I., Cahn, A. R., and Thomas, L. J. 1933. *Klebsiella paralytica* / A new pathogenic bacterium from "moose disease". Jour. Infect. Dis., 35: 386-414. Reprinted "with additions" by the Amer. Med. Ass. Press, Chicago.

Wayson, N. E. 1927. Public Health Report, U.S. Public Health Service, 42: 1489.

Wherry, W. B. and Lamb, B. H. 1914. Infection of man with *Bacterium tularense*. Jour. Infect. Dis., 15: 331-340.

Wherry, W. B. 1914a. A new bacterial disease of rodents transmissable to man. U.S. Public Health Report, 29: 3387-3390.

Wherry, W. B. 1914b. Discovery of *Bacterium tularense* in wild rabbits, and the danger of its transfer to man. Jour. Amer. Med. Ass., 63: 2041.

White, P. B. 1929. A system of bacteriology in relation to medicine, 4: 132-3.

Williams, M. Y. 1920. Notes on the fauna of the Moose river and the Mattagami and Abitibi tributaries. Can. Field-Nat., 34: 121-126.

Williams, M. Y. 1921. Notes on the fauna of Lower Pagwachuan, Lower Kenogami and Lower Albany rivers of Ontario. Can. Field-Nat., 35: 94-98.

Williams, M. Y. 1922. Biological notes along 1400 miles of the Mackenzie river system. Can. Field-Nat., 36: 61-66.

Williams, M. Y. 1933. Biological notes, covering parts of the Peace, Laird, MacKenzie and Great Bear lake basins. Can. Field-Nat., 47: 23-31.

Williams, M. Y. 1933. Fauna of the former Dominion Peace river block, British Columbia. Rept. Prov. Mus. Nat. Hist., for 1932, pp. C14-C24.

Wing, L. W. 1934a. Migration and solar cycles. Auk, 51: 302-305.

Wing, L. W. 1934b. Cycles of migration. Wilson Bull., 46: 150-156.

Wing, L. W. 1935. Wildlife cycles in relation to the sun. Trans. 21st. Amer. Game Conference, pp. 342-363.

Woodruff, L. L. 1911. The effect of culture medium contaminated with the excretion products of *Paramecium* on its rate of reproduction. Proc. Soc. Exp. Biol. and Med., 8: 100.

Wu Lien-Teh (Tuck). 1924. A further note on natural and experimental plague in Tarbagans. Jour. Hyg., 22: 329-334.

Yersin. 1894. La peste bubonique à Hong Kong. Ann. Inst. Past., 8: 662-667.

Yorke, W. and Maplestone, P. A. 1926. The nematode parasites of vertebrates. London.

www.ingramcontent.com/pod-product-compliance
Lightning Source LLC
Chambersburg PA
CBHW030526210326
41597CB00013B/1042